Nuclear Power

Scrivener Publishing
100 Cummings Center, Suite 541J
Beverly, MA 01915-6106

Publishers at Scrivener
Martin Scrivener (martin@scrivenerpublishing.com)
Phillip Carmical (pcarmical@scrivenerpublishing.com)

Nuclear Power

Policies, Practices, and the Future

Darryl Siemer

Scrivener Publishing

WILEY

This edition first published 2019 by John Wiley & Sons, Inc., 111 River Street, Hoboken, NJ 07030, USA and Scrivener Publishing LLC, 100 Cummings Center, Suite 541J, Beverly, MA 01915, USA

For more information about Scrivener publications please visit www.scrivenerpublishing.com.

Wiley Global Headquarters
111 River Street, Hoboken, NJ 07030, USA

For details of our global editorial offices, customer services, and more information about Wiley products visit us at www.wiley.com.

Library of Congress Cataloging-in-Publication Data

ISBN 978-1-119-65778-1

Cover image: purchased from Dreamstime.com
Cover design by Kris Hackerott

Set in size of 11pt and Minion Pro by Manila Typesetting Company, Makati, Philippines

10 9 8 7 6 5 4 3 2 1

Contents

Introduction

Apparently because I am among a minority of scientists [13, 23, 24, 257, 281] willing to say that today's politically correct (non-nuclear) renewable energy sources couldn't support even the near future's (2050 AD) human population without severe environmental consequences, I was asked to contribute a chapter to an upcoming soil science book describing how a "nuclear renaissance" could address Africa's especially imposing future issues [280]. That morphed into one of QUORA's longest-ever winded "answers" which, in turn, inspired this effort [274].

I'm a 74 year old, retired Idaho National Laboratory (INL) "Consulting Scientist" (first, analytical research chemist, then chemical engineer, and finally, radwaste materials scientist) who had spent much the last ten years of his career "whistleblowing" about how INL's previous mission (serving as DOE's "lead lab" in radwaste management) had been (and is still) being managed. This book will first explain why nuclear power should and could become the world's primary energy/power source and then identify the reasons that it's become so difficult to develop a "nuclear" technological fix for the future's energy conundrums. It is neither another review of the history of nuclear power nor a summary of other peoples' work and opinions. For the most part, others' work/opinions will be summarized and "open access" references given so that readers can read them themselves (which I encourage). It is largely autobiographical and seeks to encourage its readers to do the same thing that's guided me throughout my scientific career: "*think for yourself and do it with numbers*"[1]. The biggest difference between this and the books written by others sharing my enthusiasm for nuclear power (e.g., [102, 163, 164, 199, 208, 249, 253, 258, 275, 276], is I am a technical nerd "ex insider" more aware of the nuclear industry's foibles.

[1] The reason for basing decision/opinions upon "numbers" is that many ideas/claims that initially appear to be perfectly reasonable prove to be unreasonable when examined quantitatively.

Anyway, after I'd retired, I decided to take advantage of the "openness" provided by the internet (not DOE) to determine for myself whether or not we actually needed a "nuclear renaissance", the development/promotion of which had become INL's "new mission". After pondering the current suite of alternatives had convinced me that the world does indeed need such a renaissance, I then looked into why we didn't just go ahead and implement one capable of addressing the Future's energy-issues; i.e. one based upon breeder-type reactors[2]. The root cause of that failure turned out to be the same cultural "symptoms" responsible for turning the treatment of DOE Hanford's/INL's "high level wastes" (HLW), opening a HLW waste repository, & building a MOX-type[3] reactor fuel factory, into multibillion dollar boondoggles. My own experiences within DOE's laboratories along with what's happened since I retired leads me to suspect that most of its nuclear engineers and scientists quite understandably are not "really serious" about much other than continuing to quietly study their little non-controversial piece of "all of the above"[4] until they can retire too.

Today's economic development models are largely based upon continued consumption of fossil fuels and therefore pose serious threats to the environment. Local side effects include pollution, land degradation, forest fires, and water scarcity. Globally, climate change creates increased risk of extreme weather events such as floods and drought, as well as changes to our ecosystem like sea level rise, forest loss and ocean acidification. Climate change is already impacting millions of people, particularly the world's most vulnerable populations, which is why mankind really does need to embark upon an all-out effort – a "nuclear green new deal" - to combat climate change.

There is a tremendous need for more and better "energy services" to fuel economic development and provide energy security throughout the entire world, not just within its currently rich societies. By circa 2100 AD (and preferably sooner) we must build "clean" (GHG-free) energy systems

---

[2] A breeder reactor generates at least as much fissile material as it consumes. They can do this because their neutron economy is high enough (they don't waste neutrons) to create ("breed") more fissile (^{233}U, ^{235}U, or ^{239}Pu) than they "burn" (fission) via neutron capture by fertile materials (^{238}U or ^{232}Th) also within the reactor. Breeders were initially considered absolutely necessary because they make far more efficient use of natural uranium (NU) than do light water reactors, but interest plummeted after the 1960s when more uranium was discovered and a cheaper way of doing uranium enrichment (ultracentrifugation) was developed.

[3] MOX = "mixed oxides" of plutonium and uranium.

[4] During the last two administrations, "all of the above" has included gas, oil, wind, solar (two flavors), geothermal, low/high hydro, wood/palm oil/switchgrass/corn/rapeseed/soy-based biofuels, "beautiful clean coal", and nuclear power.

capable of powering the homes, factories, transportation systems, and cities in a world that's even more environmentally compromised with ~50% more people to support. The supply, speed, and scale of such change is unprecedented. Those needs cannot be satisfied for some people at the expense of others – the rich can't just keep getting richer while everyone else's lives become more precarious.

While this book will demonstrate why a properly implemented nuclear renaissance could address the future's energy-related issues more effectively than could any combination of politically correct renewable energy sources, it won't dwell exclusively upon such things – it'll be pro nuke, not anti-anything except the cultural pathologies that have rendered significant progress in this arena almost impossible in my country (USA). My goals are to first, convince my ex colleagues that it is absolutely necessary to get serious about implementing a sustainable version of such a renaissance, and second, remind them that there's a pretty good chance of succeeding if they ever become willing to pull their heads out of their leaderships' drawers. A third is to explain why the US federal government's nuclear engineering (NE) experts shouldn't be entrusted to do "the right thing" with respect to developing a practical solution to the future's energy problems. To support that contention, I will present examples based upon my own experiences with the US Department of Energy's (DOE's) approach to managing nuclear energy related projects.

Since the African continent's people are apt to be the most severely impacted by the future's demographic, environmental, resource limitation, and economic challenges, most of my specific examples will invoke its issues. Like those in David Mackay's book "Sustainable Energy: Without the Hot Air" [266], my examples will be numeric (quantitative) and based upon reasonable assumptions and readily obtained (GOOGLEable) data, not sweeping generalizations, over simplifications, or the sorts of wishful thinking often reflected in disquisitions invoking nuclear powered utopias. This will undoubtedly offend some of my pro nuclear colleagues because some of the reactor concepts that they are championing are much less "equal" than others [272].

I'll be putting the results of my numeric examples into proper perspective: "naked" numbers – especially, very large or small ones presented with "strange" units[5] – often mislead even scientists and engineers. This book's scenario is that by 2100 AD, a "sustainable nuclear renaissance" – not an "all of the above" mix of today's politically correct renewable energy sources

[5] For example, units like "barrel of oil equivalent" (BOE) are often rendered even less intuitive with unmentioned assumptions having to do with how that energy is to manifest itself (as process heat? electricity?). Little "technical details" like that are important.

or more-of-the-same nuclear reactors– will be addressing the root causes of most of mankind's misery throughout the ages. Among the things that such a renaissance would accomplish, food production would become genuinely sustainable because cheap/clean electrical power would simultaneously address the world's water woes and enable the mining, grinding, shipping, and distribution of sufficient powdered basalt over farmland to affect Mother Nature's too slow, notoriously unreliable and sometimes catastrophic (volcanic) approach to both soil-building and atmospheric CO_2 removal. It would also enable the mining, grinding, shipping, and distribution of sufficient ultramafic rock-based sand to sea shores and reefs to protect them from rising sea levels and reverse oceanic acidification [78].

The world's nuclear scientists, especially the USA's, must screw up enough courage to discard their blinders and concentrate upon solving real problems. Addressing the future's energy conundrums will require lots of big, "sustainable" reactors powering a world-wide industrial "new deal" like that envisioned by Alvin Weinberg almost a lifetime ago - if thinking that way makes me "unpatriotic", I guess I am.

1

Africa's Especially Special Issues

Unlike most first world nations, the majority of Africa's 54 countries continue to exhibit alarmingly high rates of both population growth and poverty [1]. Approximately 380 million of its ~1.2 billion people are extremely poor – often hungry – and ten of the world's most underdeveloped [2] countries – Mozambique, Guinea, Burundi, Burkina Faso, Eritrea, Sierra Leone, Chad, Central African Republic, Democratic Republic of Congo, and Niger – are located therein. Furthermore, although considered exceptionally underdeveloped, none of them are among the twenty countries recognized to possess the world's lowest living costs (Cheap 2018) meaning that Africa's poor people are considerably poorer in fact than are those in more technologically advanced but poor by OECD standards nations like Romania. Most of Africa's people are plagued by a lack of basic infrastructure due to dysfunctional and, often, self-serving governance further complicated by long festering civil/tribal/religious conflicts and therefore face bleak futures[6]. Much of Africa is also apt to be particularly hard-hit by climate change - the Sahara desert is getting bigger[7]. The fact that most of its countries are ill equipped to deal with any sort of natural disaster, possess economies comprised primarily of subsistence farming on progressively poorer-quality land, and have grossly underfunded public health, physical infrastructure, and education services constitute only some of the factors considered in compiling quality-of-life rankings. Most of the United Nation's measures of Human Development [3] also consider the fairness of income/wealth distribution for which Africa's countries are also especially low-ranking [4]. Cambridge's Sir Partha Dasgupta, recipient of almost every award that economists can bestow, has pointed out that most of the

[6] "*Life has become more brutal and brutish*", Wole Soyinka (84 year-old Nigerian playwright & philosopher). CGTN interview, 16Feb2019.

[7] Much of, Spain, Portugal and the southwestern US is becoming more desert-like and each year's heat waves are killing more people everywhere.

Darryl Siemer (ed.) Nuclear Power: Policies, Practices, and the Future, (1–6)

recent GNP increases of $2^{nd}/3^{rd}$ world countries have come at the expense of their average citizens' personal assets [5].

Africa's (and the World's) still burgeoning population growth exacerbates all of its problems. As of 2015, the UN's mid-range population growth projection is that Africa will have ~4.5 billion inhabitants by 2100 AD – about three times that similarly anticipated for the world's currently most populous nation, China. The populations of 28 African countries are predicted to more than double between 2015 and 2050 and, by 2100, those of Angola, Burundi, Democratic Republic of Congo, Malawi, Mali, Niger, Somalia, Uganda, United Republic of Tanzania and Zambia are to increase at least five-fold.

Frankly, I consider such projections unrealistic for the following reasons. First, the world's current political trends - increasing "populism" (extreme polarization often bordering upon fascism) driven primarily by rapidly increasing class, power, and wealth disparities. Second, the armed-to-the teeth "leader of the western world's" nervousness about the fact that its dominance of the world's economic system is rapidly diminishing. Third, additional population stressors driven by climate change and resource limitations. Fourth, and finally, most of human history (nature?) suggests that those factors could ignite another "world war" apt to kill far more people than did the 20th century's and thereby curb (or reverse) population growth.

Unless a new, worldwide, "Fair Deal" somehow comes to pass[8], the relative demographic weight of the world's developed countries will drop shifting economic power to developing nations. The already-developed countries' labor forces will age and decline constraining their economic growth and raise the demand for "cheap" non-documented immigrant workers which will likely further increase the frequency of killings, burnings, and bombings driven by jingoistic populism. Most of the world's population growth will be concentrated in the poorest, youngest, and most heavily faith-based (mostly Muslim) countries many of which will continue to be unable to provide adequate education, capital, and employment opportunities for their young people.

Finally, most of the world's population will live in cities, with the largest such heavily urbanized areas in the poorest countries, where adequate

---

[8] The Fair Deal was an ambitious set of proposals continuing Roosevelt's New Deal liberalism presented by U.S. President Harry S. Truman in his 1949 State of the Union address. It's most important proposals were aid to education, universal health insurance, the Fair Employment Practices Commission, and repeal of the Taft–Hartley Act. However, because a Conservative Coalition controlled Congress, they were all debated at length and then voted down.

policing, sanitation, health care, and even clean water are likely to be available only to their richest inhabitants. Such urbanization is apt to be profoundly destabilizing. People moving to cities within developing countries during the rest of this century are apt to have far lower per capita incomes than did those of most of today's industrial countries when they did so. The United States, did not reach 65 percent urbanization until 1950, when its per capita income was nearly $118,000 in 2019 dollars. By contrast, countries like Nigeria, Pakistan, and the Philippines now approaching similar levels of urbanization, have per capita incomes of $2,300–$5,200. Countries with younger populations are especially prone to civil unrest and less able to create or sustain democratic institutions. The more heavily urbanized they become, the more they are apt to experience grinding poverty and anarchic violence. In good times, a thriving economy might keep urban residents employed and governments flush with sufficient resources to meet their needs. More often however, people living in sprawling, impoverished cities are victimized by crime lords, gangs, and petty rebellions. Thus, the rapid urbanization of the developing world is apt to bring in more exaggerated form, the same problems that urbanization brought to nineteenth-century Europe: cyclical employment, inadequate policing, and limited sanitation and education which spawned wide-spread labor strife, periodic violence, and sometimes, even revolutions. International terrorism originates in fast-urbanizing developing countries. Within poor sprawling megacities like Mogadishu and Damascus, neighborhood gangs armed with Internet enabled social networking offer excellent opportunities for the recruitment, maintenance, and hiding of terrorist networks [6].

When life is cheap, worthwhile jobs unavailable, and the future looks even worse, young people often to go to war.

US Pentagon studies [7] concluded that the root cause of majority of such deaths will be disease and starvation engendered by the disintegration of technology-dependent societies dependent upon increasingly limited/degraded resources (land, food, fuel, high grade ores, etc). In his book, "Small is Beautiful: A Study of Economics as if People Mattered", Ernst Schumacher [8] observed that today's technological civilization is unsustainable because the finite resources enabling it are treated as inventory (income) rather than capital. The sustainability of today's economic systems therefore requires continued growth of both population[9] and total wealth (GDP), both of which are impossible in a finite world.

---

[9] The Earth now supports about three times as many people as it did when I was born and five times more than when my grandfather was. Most of that growth is due to the fact that our energy-enabled civilization's technological advances have decreased child mortality

Second, if our world's future leaders were to decide to implement the changes needed to address the technical issues otherwise likely to lead to war, doing so would result in much lessened fertility [9]. Figure 1 depicts the effect that increasing people's prosperity[10] has upon their reproductive choices (replacement fertility ~2.1) - if the future were to become both much richer and fairer than it is now, today's unsustainable population growth would quickly end.

Consequently our leadership's objective should become encouraging the development of a genuinely sustainable and much more egalitarian world in which every individual regardless of where they live or who they are does indeed matter[11]. Until they acknowledge the above-related facts, embrace appropriate goals, and begin to act accordingly, we'll just continue to spin our wheels while blaming "the other guys".

(not raised birth rates) and rendered it possible to feed far more people. I suspect that if this book's scenario were to come to pass, human population will gradually drop back to a level (2-3 billion?) consistent with both much more comfortable lives for individuals and more room for living creatures other than us.

[10] Norway's performance at the last Winter Olympics exemplifies how the "UN's Inequality adjusted human development index" reflects a county's citizens' "quality of life". Half way through that gathering, Norway, with only 1.6% of the USA's population, had won almost twice (37/21) as many medals. Why? It's not because its people are "richer" than are the USA's: a bit of GOOGLING reveals that Norway's after-tax GDP/capita income level is almost identical to the USA's - $45,348 vs $45,648. The real reason is that because Norway's people govern themselves in a way that benefits them rather than powerful special interests including overly-entrenched duopolistic political parties – their tax dollars support them, not serve special interests. Its policies have generated a much more equitable distribution of wealth/GDP top-to-bottom than the USA has (it's got very few poor and almost no homeless people); its medical service providers can't force anyone into bankruptcy due to injury or illness; its public schools are much better than the USA's; its federal government doesn't profit by increasing its college students' debt burdens; and its employers must grant its citizens far more paid leave than the USA's are entitled to (25 days/year Norway vs zero USA). It's also unlikely that Norway's government routinely forces its "essential" employees to work without pay (slavery?) while its politicians bicker with each other. Consequently, Norway's people feel much more secure, freeing them to do whatever they wish, including fun things like skiing that only "lucky" Americans can afford. This argument explains the relative per capita athletic performance of other northern European nations' and Canada's Olympic teams to the USA's. The USA's electorate also doesn't seem to care that their top political leadership's credo seems to be: "*For whosoever hath, to him shall be given, and he shall have more abundance: but whosoever hath not, from him shall be taken away even that he hath*" (Matthew, 13:12).

[11] "The test of our progress is not whether we add more to the abundance of those who have much; it is whether we provide enough for those who have too little." — Franklin D. Roosevelt

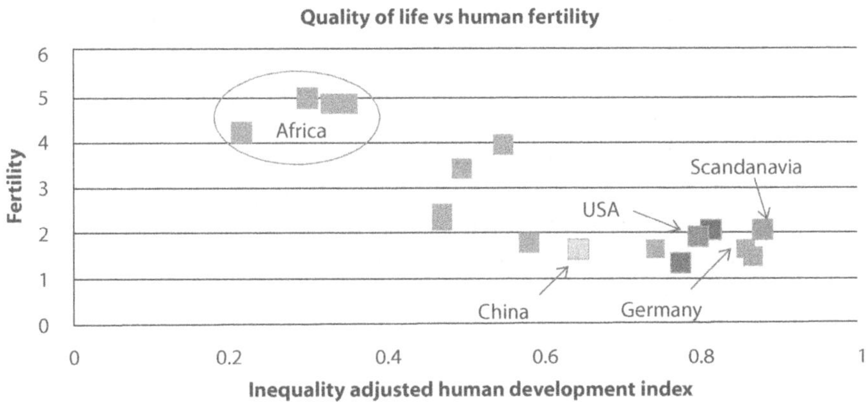

Figure 1 Prosperity vs human reproductive choices (WIKIPEDIA data).

What are those goals?

Since food represents any living creature's most fundamental need and its source for humanity is farm land, I'll begin by describing what's been happening along those lines in the world's currently most undeveloped continent, Africa. A recent Brookings Institute report [10], points out that, "*no matter how effectively other conditions are remedied, per capita food production in Africa will continue to decrease unless soil fertility depletion is effectively addressed.*" It goes on to say that a second major problem with the oft-assumed African "land abundance" hypothesis is its inconsistency with convincing evidence that its soils are being simultaneously depleted and eroded by current agricultural practices including a decline in fallowing. While some African leaders along with the management of "land grabbing" (?) international agribusiness concerns seem to feel that Africa still has plenty of yet-undeveloped arable land, many of Africa's poorest people (mostly subsistence farmers) can't afford to let any of theirs lie fallow and thereby recover: some families live on 0.36 ha (0.9 US acre) farms yielding under 1 t of grain/ha ($t = \text{tonne} = 10^{+3}$ kg $= 10^{+6}$ grams: $\text{ha} = \text{hectare} = 10^{+4}\ m^2 = 2.59$ US acre) while the first-world's farmers routinely produce 3 to 12 t/ha of whatever cash crop they chose to plant on several order of magnitude larger farms.

The key differences between the agricultural practices of developed nations and most of Africa's include:

1. Developed nations heavily fertilize their croplands – most of Africa's farmers can't afford artificial fertilizers and often have to burn any manures or crop residues they can gather to cook their food.

2. Developed nations' farmers can afford to irrigate their croplands – most of Africa's can't. That issue is compounded by the fact that much of Africa's nominally arable land doesn't get enough rain to reliably support anything other than skeletal cow or goat grazing.
3. Most developed-nation farms are both large and productive enough to enable their owners to buy/utilize specialized machinery which renders their labor far less exhausting and much more rewarding[12]. The world's poorest farmers still work themselves to death with primitive tools.

[12] Including its little "hobby farms", the average size of farms in the USA's Corn Belt is about 350 acres and such land is now worth about $9000/acre.

2

Why Everything Boils Down to Energy Inputs

It is evident that the fortunes of the world's human population, for better or for worse, are inextricably interrelated with the use that is made of energy resources.

M. King Hubbert

The differences simply reflect the relative amounts of raw/primary energy supporting the lifestyles of rich vs. poor people, which still boils down to their relative per capita fossil fuel consumption. Since I'm going to be supporting most of my contentions with the results of ball park calculations, let's start this off with a table containing many of the numbers/terms that will be used throughout complete with their units (for brevity's sake I'll be leaving the units out of most of this book's example calculations).

On-farm agricultural energy consumption in rich (developed) countries entails the burning of diesel oil, gasoline, and/or LP gas in engines plus use of electricity made by burning another fossil fuel, usually coal. Consumption is considerably higher in high-GDP countries (around 20.4 GJ/ha) than it is in low-GDP countries (around 11.1 GJ/ha) and far greater than on Africa's subsistence farms [11]. For example, most of the energy input to a subsistence farm consists of human labor which, throughout an 8 hour work day, averages about 75 watts per person [12]. If 100% of such useful (in this case, mechanical) energy [2.16E+6 J/day = 75*8*3600] is devoted to cultivating a 0.9 acre (0.36 ha) plot throughout a 6 month growing season, the area-normalized "energy services" devoted to it is 1.08 GJ/ha/a [2.16E+6 J/a*365days*6/12/2.47 acre/ha/1E+9], which is about ten percent of the raw/primary food energy required to keep each person so-occupied alive throughout the entire year [2500 kcal/day *365 days/a]. Energy-wise that's not very efficient – state of the art farm machinery generates about one third of a joule's worth of useful energy from a joule's worth of fuel heat energy and doesn't consume anything when not running.

Darryl Siemer (ed.) Nuclear Power: Policies, Practices, and the Future, (7–34)

Table 1 "Special" numbers, constants, etc. along with their units.

3600	= 3600 s/hr
24	= 24 hrs/day
365	= 365 days/year = 365 days/a
3.2E-11 (energy)	(energy) = # of Joules generated by the fission of a single actinide atom (='s 200 million electron volts (MeV)
6.023E+23	= 6.023E+23 = number of atoms, molecules, etc., per gram mole of anything
1.6E-19	= Number of electron volts per Joule (the combustion of a single carbon atom generates 4.1 electron volts)
Watt (power)	(power) = W = energy/second = Joule/s = J/second = J/s = 0.00134 mechanical horsepower
kWh	(energy) kilowatt hour = 1 J/s*3600 s/hr*1000 = 3.6E+6 Joules (most common unit for electrical energy)
GWyear (energy)	= Giga $W_x$yr = 1E+9*3600*24*365 = 3.15E+16 J (x) may be thermal (x=t) or electrical, x=e (GWe)
Calorie (energy)	(energy) = 4.19 J = 0.001 kilocalorie =0.001 kcal
Acre (area)	= 43560 ft^2 = 4049 m^2 = 0.4049 ha = 0.0015625 $mile^2$
BTU (energy)	= British thermal unit =1055 J (energy required to heat one pound (454 g of water) 1 degree F or 5/9 degree C)
Quad	(energy)= one quadrillion BTU = 1.055E+18 J
BOE	(energy) = barrel of oil equivalent=6.1E+9 J (assumes a 42 US gallon barrel, 10 kcal/g, and 0.916 g/cc oil)
nX	= 10^{-9}*X = nano X
kX	= 10^{+3}*X = kilo X
MX	= 10^{+6}*X = mega X
GX	= 10^{+9}*X= Giga X
TX	= 10^{+12}*X=Terra X
EX	= 10^{+18}*X= Exa X

According to a "Food and Agriculture Organization of the United Nations" (FAO) report [13], the raw/primary energy consumed by the world's "food sector" amounts to ~95 EJ (exa ($10E+^{18}$) Joules) – approximately 20 percent of current total global raw/primary energy consumption (~570 EJ) – and generates over 20 percent of anthropogenic greenhouse gas emissions. Land use changes, particularly those linked to the deforestation brought about by the expansion of agricultural lands to raise food crops and biofuels [14] constitutes another ~15 percent of anthropogenic GHG emissions. Only about 5% of that energy, ~6 EJ, directly supports on-farm activities such as cultivating and harvesting crops, pumping water, housing livestock, heating protected crops, drying and, short term storage. The rest/majority of the agricultural sector's energy demand is devoted to transport, fertilizer and pesticide production, food processing, packaging, storage, and distribution.

All of the world's developed countries adopted Dr. Borlaug's fossil-fueled "Green Revolution" which enabled ~90% of today's ~7.5 billion people to consume as much food – both basic necessities along with some luxury items – as they want [15]. Approximately one half of the world's current population would quickly starve if that hadn't happened. However, the fossil fuels enabling both it and the 20th century's industrial and information revolutions generated huge environmental impacts including the greenhouse gas (GHG) emissions responsible for global warming/climate change. From 1870 to the present, fossil fuel burning dumped about 580 Gt (Giga tonnes = 580 Pg(~ 0.5 A/cm^2) C) into the atmosphere in the form of ~2100 Gt of CO_2. That gas partitioned between the atmosphere, oceans and land, warming all of them and thereby causing increasingly severe and frequent weather events including "Super El Niños" [16], ocean acidification, drought and biofuel production-driven food cost escalation, air pollution, deforestation, potable/irrigation water shortages, sea-shoreline erosion/flooding, and relentless cost of living increases in the world's poorer regions. Those effects constitute threat multipliers that aggravate human stressors – poverty, environmental degradation, hunger, political instability, and social tensions – and thereby engender mass migrations plus a great deal of terrorist activity and other forms of violence.

James Hansen probably possesses the world's most "educated" opinions about the causes, effects, and consequences of global warming. They've been summarized as follows [17]:

1. Climate has always changed, but humans are now the main driver for change
   a. Rising atmospheric CO_2 levels, primarily a result of fossil fuel emissions, have become the predominant cause of continuing climate change

   b. Climate change is driven by cumulative CO_2 emissions. The U.S. has contributed a disproportionately large share of cumulative global emissions.
2. Current levels of atmospheric greenhouse gases (GHGs), mainly CO_2, cause Earth to be out of energy balance. This imbalance is driving climate change.
   a. Earth's energy imbalance is now measured and large. As long as Earth remains out of energy balance, the planet will continue to get hotter.
   b. If GHG amounts continue to rise unabated, the energy imbalance will drive global warming to levels with climate impacts beyond the pale (see 3).
3. If high fossil fuel emissions continue unabated, consequences will be predominantly negative for humanity, especially for young people.
   a. Sea level: Continued high fossil fuel emissions will eventually make coastal cities dysfunctional, with incalculable consequences.
   b. Species exterminations: Shifting of climate zones, with other stresses, may commit many species to extinction, leaving a more desolate planet.
   c. Regional climate: subtropics and tropics will become dangerously hot, if high emissions continue. Emigration chaos may threaten global governance.
4. Required actions to avoid dangerous climate change are guided by Earth's climate history and by the need to restore Earth's energy balance
   a. Science can specify initial targets, sufficient to define policy needs
   b. Emission reductions must begin promptly, or climate will be pushed beyond a point at which changes proceed out of human control
5. The U.S. government, via both actions and inactions, is behaving with flagrant disregard of rights and well-being of the public, especially young people
   a. Action: authorizing, permitting, subsidizing massive fossil fuel extraction
   b. Inaction: absence of any coherent, effective program to reduce them

During the ~150 years since we began to power ourselves with fossil fuels, two world wars and numerous smaller ones have been fought over access to natural resources - primarily "lebensraum" (land) and fossil energy resources, usually petroleum - some of which resulted in the Christian-country "winners" creating new countries in the oil rich Islamic Persian Gulf, which, of course, eventually engendered more conflict.

Securing those resources has proven to be expensive to those war's winners. A Princeton University report concluded that simply keeping the US Navy's fifth fleet within the Persian gulf from 1976 to 2006 had cost its taxpayers ~$6.8 trillion 2008 dollars and would probably cost them another $0.5 trillion during 2007 [18] which figures didn't include the costs of actual conflicts. Since that fleet remains on station, the total cost of "maintaining presence" therein has now probably reached about $12 trillion. A 2013 Kennedy School of Government report [19] concluded that the total cost of the USA's most recent wars in the Middle East and Northern Africa would probably be $4-6 trillion and had accounted for roughly 20 percent of its increase in national debt between 2001 and 2012 (modern wars are fought with borrowed money).

Concerted international effort to address fossil fuel's environmental impacts began with the UN's 1997 Kyoto Protocol to which many, mostly small and not particularly impactful, countries signed up. While the billions of dollars spent on climate science research since then have generated thousands of papers/reports and paid for hundreds of other conferences both large and small, neither that science nor the policy changes of many countries favoring/subsidizing politically correct renewable energy have had much effect upon mankind's GHG emissions. As this is being written (December 2018) representatives from 195 countries have again gathered (in Katowice, Poland) for this year's United Nations Climate Change Conference, COP 24 (COP = "Conference Of the Parties"). This meeting is focused upon producing rules to flesh out the "details" of the 2015 Paris Climate Accord (COP 21), the landmark agreement signed by all of its attendees except Nicaragua and Syria, to battle climate change and, hopefully, limit global warming to 1.5 degree Celsius, one-half degree under the 2°C limit set earlier at COP 15 (the Copenhagen conference). Since 2015, the International Panel on Climate Change's (IPCC's) leadership has been trying to breathe new life into that accord amid backsliding from several key nations, most notably the United States, over commitments made when they signed it. To date, the IPCC's efforts have not really accomplished much because key "parties" refuse to agree upon a mechanism ensuring that they honor their commitments with respect to either GHG emissions or contributions to a $100 billion/a climate mitigation fund.

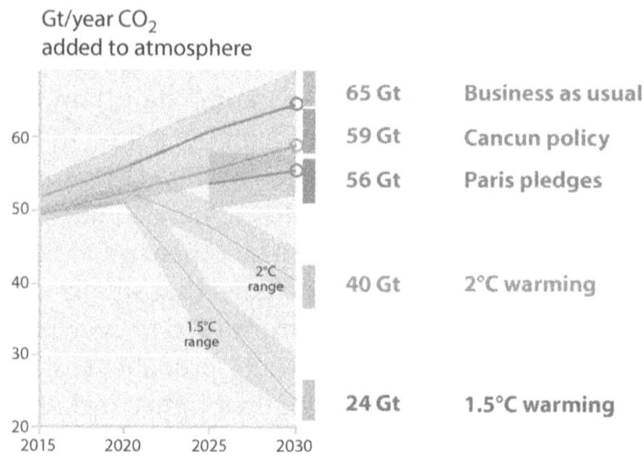

Figure 2 Mankind's CO_2 equivalent[14] "Emissions Gap" (Source: UN IPCC 2018 Emissions Gap).

The latest version of the British Petroleum company's annual Statistical Review of World Energy [20] contains not only information from the preceding year, but also historic data on consumption and production of all forms of energy during the last several decades. Its principal conclusion is that humanity is not reaching the goals established by the Paris Agreement (see Fig. 2). In 2017, Mankind took a step backwards with respect to the timid advances made during the two preceding years: the use of fossil fuels had grown, increasing CO_2 emissions by ~1.6%. That trend continues – anthropogenic CO_2 emissions rose another 2.7% in 2018 [21]. Worse, most climate models indicate that by 2100 AD, even if the emission "commitments" made by COP 21-24's attendees were to be honored, they would likely cause global warming of between 2.7 and 3.2 degrees Celsius, well above the 1.5–2 degree threshold that many climate modeling experts consider a tipping point beyond which Nature's positive (?) feedback mechanisms[13] will render catastrophic impact inevitable [23, 24].

[13] For instance, a recent paper in Nature [22] indicates that for each degree that the Earth's temperature rises, the amount of methane entering its atmosphere from microorganisms dwelling in lake sediment and freshwater wetlands -- the primary sources of that gas -- will increase several fold. As temperatures rise, the relative increase of methane emissions will outpace that of carbon dioxide from those sources. There's also vast amounts of methane trapped within the Arctic's currently frozen muskeg, which, along with that emitted by more southern wetlands and ricefields, is apt to cause runaway global warming.

[14] "CO_2 equivalent" is a measure of all anthropogenic GHG emissions in terms of the amount of pure CO_2 required to have the same effect. CO_2 itself currently represents ~75% of the total.

The most important thing that we must do is to quickly replace today's fossil fuel based energy system with something that is both "clean" (no GHG gas emissions) and sufficiently reliable (not intermittent) to power a bigger, more interconnected, more prosperous, cleaner, fairer and happier world than is the one we're living in today.

> *"To prevent the worst effects of climate change, we need to reach near-zero emissions on all the things that drive it—agriculture, electricity, manufacturing, transportation, and buildings—by investing in innovation across all sectors while deploying low cost renewables," Nuclear energy is one of these critical technologies. It's ideal for dealing with climate change, because it is the only carbon-free, scalable energy source that's available 24 hours a day."*
>
> **Bill Gates**

2.1 How Much Clean Sustainable Energy Will Our Descendants Need?

The exceptionally "rich" lifestyle[15] of the USA's ~320 million people is nominally supported by about 99.5 EJ (98 quads) of raw/primary energy per annum, which figure has remained roughly constant for over two decades [25]. It's "nominal" because its consumer-driven economy consumes energy and other resources from an area outside of and greater than the USA's which isn't counted in such compilations. "Ecological footprint analysis" [26] provides us with a more realistic measure of the USA's resource consumption. Table 2 provides those figures for 2009 – about 41% of the USA's GHG emissions were incurred in/by other countries.

Currently, all of mankind consumes ~2,300 GW's worth of electrical power which works out to about 307 watts/person. However, like most of the things that determine human life styles, its distribution is extremely uneven. People in Scandinavian countries consume the most, about 2500 watts per person, followed by the USA and Canada's ~1400 W. However, the majority of Latin America's people consume less than 250 W, South Asia's below 100 W, and Africa's, under 25 W. Over a billion people now don't have access to electricity at all. If mankind is to prosper, clean,

[15] "rich" in terms of material things – most of the USA's people own more stuff", drive further in bigger cars and live in much larger houses than do average Europeans.

Table 2 US Greenhouse gas emissions including those originating from products/services made abroad.

Service	Percent total GHGs
Infrastructure	1
Appliances & devices	7
Non local passenger transport	9
Food provision	12
Local passenger transport	13
Building HVAC & lighting	21
Provision of other goods	37

(*from* Joshuah Stolaro, Products, Packaging and U.S. Greenhouse Gas Emissions, Athens, GA, Product Policy Institute, September 2009).

affordable and dependable (not intermittent) power must become available to everyone and it must be provided in a way that doesn't pollute the air, poison the land, or change the climate.

Anyway, ~80 percent of the USA's primary/raw energy is provided by fossil fuels translating to a mean per capita raw/primary energy consumption rate (power) of 9860 watts [99.5E+18J/3600/24/365/320E+6] or about eight [99.5/3.2e+8/570/7.5E+9] times that of the world's average person today. Since one joule's worth of raw/primary (heat) energy currently provides about 0.4 joules worth of useful "energy services" (the efficiency of most of fossil fuel's applications is Carnot-limited) and Europeans apparently live almost as well consuming one-half that much raw/primary energy per capita, let's assume that supporting the life styles of each of the future's equally EU-rich people would require ~2 kW's [9860* 0.5*0.4 = 1972 ≈ 2000] worth of energy services (electricity). Consequently, a world with 11.2 billion such people must possess power plants able to supply an average power of about twenty two [11.2E+9*2000*3600*24*365/1E+12/3600/24/365 = 22.4] TW_e (terawatt electrical). Finally, assuming that each individual region's peak power demand is ~40% higher than its average and that no world-wide, zero-loss, "super grid" exists, our descendants would need ~30,000 [22.4*1.4*1012/109] one GWe power plants to live that well.

That power could not be generated with fossil fuel because even if there were enough of it (there isn't[16]), burning it would have catastrophic consequences. For example, the raw/primary (heat) energy represented by the world's remaining 1139 billion tonnes of coal reserves, 187 trillion m^3 of natural gas, and 1.707 trillion barrels of petroleum, [28] is about 5.0E+22 J's which, if consumed by 40% Carnot efficient power plants, could generate 22 TWe for 29 years - ~35% of a typical first-world human life span. Additionally, those reserves collectively contain about 1200 Gt of carbon which, if converted to CO_2 and dumped into the atmosphere would push global warming well past any of the "tipping points" suggested by the world's climate modeling experts.

In order to more clearly see what these facts and figures mean, it's necessary to consider them on a longer time scale than that which we humans customarily employ. Figure 3 was excerpted from a paper written/delivered by the one of the petroleum industry's most influential geologists (and, eventually, most influential gadfly), Professor M. King Hubbert, sixty three years ago [29]. It depicts Mankind's total energy consumption extending from the dawn of recorded history 5000 years ago to 5000 years in the future assuming that population eventually stabilizes and we choose to replace finite fossil fuels with a sustainable (breeder reactor-based) nuclear fuel cycle before our civilization collapses. To Professor Hubbert, "*on such a time scale, the discovery, exploitation, and exhaustion of the Earth's fossil fuels will constitute an ephemeral event.*"

Unfortunately, because the world's decision makers have been kicking the nuclear can on down the road for far too long, we'll have to continue to seek, extract, and burn fossil fuels throughout much of the rest of this century. Of them, natural gas represents the best choice because it is relatively clean burning (little smoke), generates only about one half as much CO_2/Joule as coal, and is still in a relatively early phase of depletion. According to the German Federal Institute for Geosciences and Natural Resources,

[16] We often see headlines announcing things like, "*the USA has 7 to 9 trillion barrels of oil*" which are apparently meant to reassure us that there's nothing to worry about [27]. However, if we bother to read beyond those headlines we soon discover that such oil is very "tight" and would be extremely difficult (expensive) to recover; meaning that perhaps one trillion barrels of it would be recoverable with today's fracking technologies. However, we're next assured that with "*aggressive use of new fracking technologies combined with in situ 'fire flooding' and/or 'water flooding'*" perhaps 20-30% of it might be recovered. One trillion barrels of oil represents about 610 Exa Joules worth of raw heat energy which is equivalent to about one year's worth of Mankind's current, not a richer and more egalitarian future's, energy demand.

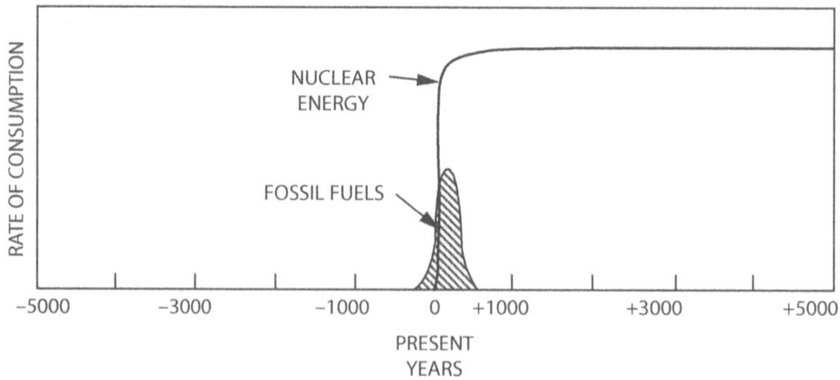

Figure 3 Mankind's long term energy consumption [29].

world cumulative natural gas production up to 2016 was 117 trillion cubic meters, world natural gas reserves were 197 trillion cubic meters, and resources were 643 trillion cubic meters ([284], Table A-15). "Resources" is defined as "*proven but which cannot currently be exploited for technical and/ or economic reasons plus unproven but geologically possible resources which may be exploitable in the future*" [30]. The 197 trillion cubic meters of gas that we know for sure(?) could be recovered with current technologies ("reserves") represents 7.23E+21 Joules of heat energy which could produce about one half that much electricity. One year's worth of 22 TW_e electrical power equates to 6.94E+20 J meaning that the world's total gas reserves could power 11.2 billion EU-energy-rich people for 5.2 years while its burning kicks the atmosphere's CO_2 concentration up by another ~50 ppm.

Because the half-life of CO_2 already in the atmosphere is about a half century [31], achieving the goals of the Paris climate accord (limiting maximum temperature rise to 1.5 °C) at this late point in time would require an almost immediate switch to clean (no GHG emissions - see Fig. 2) energy sources and enough carbon dioxide removal (CDR) from that already in the atmosphere to reduce its concentration to ~350 ppm by volume [23]. Consequently, some of the ICPP's more optimistic post-COP 20 scenarios/reports assume that "bio-energy with carbon capture and storage" (BECCS) represents a magic bullet that could address both global warming and the future's energy supply conundrum in a politically correct (no nuclear) fashion [32]. All such scenarios are unrealistic because raising sufficient switch grass, palm oil, wood, etc. would require vast amounts of land, water, and fertilizer that most people (especially the hungrier ones) would consider better-utilized if applied to producing something other than fuel. It is also unrealistic because carbon capture and sequestration (CCS) is intrinsically both difficult and expensive, which is why after several decades and many billions of dollars-worth of "study" and "demonstrations", only about 0.08% of anthropogenic CO_2 is currently being so managed [33]. Finally, BECCS-based "save the world" scenarios are impossible because they don't scale. For example, burning 100% of the world's current annual grain (about 2.5 Gt, see [34]) plus "bone dry wood" (about 1.9 Gt, see [35]) harvests in "clean" (CCS equipped) 40% efficient (optimistic) heat-to-electricity power plants would generate useful energy services (electricity) equivalent to the output of ~935 one GW_e ("full sized") nuclear reactors. That's just ~3% of the number required to render 11.2 billion people one-half as energy rich as the USA's citizens are now. Any backup system for low capacity factor energy sources (solar and wind) must be able to satisfy most, not 3%, of total demand. Furthermore, the carbon (about 1.8 Gt) in that much biofuel (primarily carbohydrate ≈ $(CH_2O)_n$, heat of combustion ~15.5E+3 J/g) represents only about 0.3% of mankind's total anthropogenic carbon emissions to date, which means that even if 100% of the CO_2 it represents were to be captured and sequestered, it wouldn't make much difference. Consequently, because it could not achieve either of the IPCC's goals and would surely compete with food, fiber, and construction-type wood production, all rosy primarily BECCS-based scenarios are hopelessly unrealistic and therefore do not "deserve further study" – the conclusion common to the majority of reports in almost every scientific field. Additionally, because growing biofuels removes inorganic nutrients and soil organic carbon, it's just another extractive technology that would further degrade the environment while compromising food production [36].

In my opinion, the most useful outcome of the climate science research performed to date is that global warming has been absolutely proven to be man-caused and that reasonably consistent/accurate estimates of global carbon fluxes, sources, sinks, etc., have joined the tremendous amount of other technical information freely available on the internet. Such information along with readily available computerized spreadsheets rendered it simple for anyone to evaluate any proposal described in a properly written/edited paper and thereby decide whether or not it is reasonable. I haven't yet seen a politically correct peer-reviewed geoengineering proposal capable of passing such muster.

For example, GOOGLEing "oceanic acidification mitigation" brings up several fine-sounding electrochemical-based schemes published in peer reviewed journals and subsequently described in press releases. A typical example invokes giant chlor-alkai cells which would electrolyze aqueous solutions of pure NaCl (natural seawater's other components would plug up such cells) to generate sodium hydroxide that would then either be dumped directly into the ocean to counteract CO_2 engendered acidification or utilized in gas/liquid contactors to scrub it from the atmosphere ([37] - see reactions below). The simultaneously produced hydrogen and chlorine gases would be recombined by fuel cells to recover some of the electrical energy required by the chlor-alkali cells

Electrolysis: $2NaCl + 2H_2O \rightarrow 2NaOH + Cl_2 + H_2$
Air scrubbing: NaOH +water+ CO_2 in air → $NaHCO_3$ aq
Energy recovery: H_2+Cl_2 (fuel cell) → 2HCl (in a water-based electrolyte)
HCl_{aq}+ a Mg/Ca-containing rock powder → $CaCl_2$ + $MgCl_2$+ rock sludge.

The fuel cells' product, HCl (a strong acid), would then be neutralized via reaction with powdered mafic (basic) rock (e.g., basalt) in giant "pressure cookers" thereby generating a waste stream comprised of decomposed rock (mostly silica) slurried-up with a magnesium/calcium/iron/ etc., chloride-salt brine. Of the numerous "technical issues" raised by such proposals, I will just discuss their electrical energy demand. While this particular example was characterized as "energetically feasible", real chlor-alkali cells require about 3.9 volts to operate at a reasonably productive rate (~ 0.5 A/cm^2) and real H_2/Cl_2 fuel cells generate only about one volt at similarly realistic current densities. This means that the net energy required to produce one mole (or equivalent) of hydroxide would be 2.8E+5 J [1 equivalent*(3.9-1) volts * 96,500 coulombs/equivalent)*1J/(or volt*coulomb)]. Producing sufficient sodium hydroxide to deal with the amount of anthropogenic CO_2 that some of IPCC's analysts apparently assumed could/would be sequestered via BECSS circa 2050 (~10 Gt/year) would require 6.36E+19J [2.8E+5 J/mole*(10E+9 t*1E+6 g/t)/44 g/mole]. If it is to be done within one year, the entire output of either ~2122 [6.36E+20 J/1 E +9 J/s (Watt)/3600 s/hr/24 hr/day/365 day/year/0.95] full-sized (~1 GW_e 0.95 capacity factor (CF) nuclear reactors, ~4.5 million 1.5 MW-rated 30% CF, wind turbines, or ~21 billion, 1 kW-rated, 10 % CF solar panels would be required.

Such schemes could not be powered with fossil fuels either. For example, since the heat of combustion of average US coal is about 24,000 J/g and burning one gram of it generates about 2.7 grams of CO_2, generating sufficient electricity to implement the above example with 50% thermal-to-electricity efficient coal fired power plants would generate about 14 Gt of "new" CO_2 – 42% more than their power could sequester in that fashion.

Another well publicized electrochemical save-the-world scheme invoked scrubbing intrinsically acidic CO_2 from the atmosphere with a strongly basic ~750°C Li_2CO_3/Li_2O molten salt electrolyte/adsorbent from which that carbon would be then electroplated-out/sequestered in the form of graphite [38]. Since both the electricity required to reduce carbonate carbon to graphite and the heat needed to keep the electrolyte molten is to be provided with "solar towers", it is/was eminently politically correct and therefore received a great deal of favorable mention. Unfortunately, because: 1) STEP's ("a solar chemical process to end anthropogenic global warming") sequestration mechanism requires four times as many electrons per carbon atom as does that of the above-described electrochemical proposal; and 2) scrubbing air with a molten salt would heat it to the latter's temperature[17], its total energy requirement would be ~four times higher *if* 90% of its process heat requirement could be recovered/recycled via heat exchangers and 19 times greater if it could not. If powered by the wind rather than solar towers, the latter figure corresponds to ~89 million 1.5 MW, 30% CF, wind turbines.

Schemes like those do not deserve "further study" regardless of who proposes them or how many warm and fuzzy renewable buttons they push.

The most alarming thing about how things have been going recently (see the Fourth National Climate Assessment - [39]) is that civilization remains absolutely dependent upon resources that will inevitably become prohibitively expensive when most of the cheaper/easier-to-access coal, oil, and natural gas have been consumed, which situation is likely to occur well before 2100 AD. Unless the world's decision makers have already developed/implemented a simultaneously "clean", reliable (not intermittent), and affordable alternative by then, civilization is apt to collapse, heralding the onset of a dark ages akin to that depicted in Mad Max movies.

2.2 This Book's Technological Fix's Specifics

Let's do some more ball park calculations to demonstrate how Weinberg and Goeller's vision could address some of Africa's (and the world's) energy-related issues.

---

[17] The heat capacity of air is ~1.05 J/g/degree. Consequently, the scrubbing of 10 Gt of CO_2 from 400 ppmv air within one year would require heating ~1.37E+12 tonnes of air from ambient to ~750°C requiring ~1.05E+21 J of energy which figure corresponds to the full-time output of 33,300 one GW_e nuclear (or methane or coal or wood chip or switch grass…-fired) power plants.

First, let's make some assumptions. To begin with, I'm (reluctantly) going to assume that the UN's population projection for what's apt to continue to be the world's most needful region (Africa) –about 4.5 billion by 2100 AD – turns out to be right.

Next, since my example will demonstrate what a nuclear renaissance should be able to accomplish with respect to assuring Africa's (and also the rest of the future world's) food security, I'm going to assume that part of the useful energy (electricity) it would provide is devoted to doing so – in other words, nuclear powered machinery would provide the water, fertilizer, and soil-building minerals required to render African agriculture sustainable.

Finally, I'm going to assume that Africa's decision makers along with who/whatever else chooses to help/enable them decide that its citizens should enjoy the same living standards as do average EU citizens today[18].

2.3 The Reasons Why Politically Correct Renewables Couldn't Save The World

Assuming that "energy services" means electricity (reasonable because most of its applications are nearly 100% efficient) and that no worldwide, zero loss, (magic) power grid has been built, generating Africa's share of the world's total energy/power supply would require about 12,000 [30*4.5/11.2*1012/109] full-sized (~one GWe) power plants generating an average output of 9 TW_e [4.5E+9*2 kW].

The following example demonstrates why today's most popular alternative/renewable energy sources could not meet that demand.

First, let's see how many of today's purportedly "cheap", roof top-type, solar panels would be required to produce future Africa's 9 TW_e's worth of useful energy. At Home Depot (Dec. 2018) one could purchase four, real state of the art (19% efficient), 265 watt-rated, 1.61 m^3 (39" by 65") solar panels for $1412. If they were to be employed in Nigeria, which purportedly exhibits an average solar irradiance of 5.5 $kWh/m^2/day$ [40], each of those panels would generate a time-averaged power of 70.1 Watts [1.61*0.19*5500*3600/(24*3600)] which means that their "capacity factor" (CF) when so situated would be 26.4% [70.1/265]. That CF is about two and one-half times greater than if such panels were installed in northern

[18] "*The test of our progress is not whether we add more to the abundance of those who have much; it is whether we provide enough for those who have too little.*" Franklin D. Roosevelt

Europe and ~50% greater than that anticipated by South Africa's energy experts circa 2050 (Table 3).

Anyway, that CF suggests that powering Africa's 4.5 billion future citizens with them would require 128 billion [9000E+9/70.1] such panels costing about $45 trillion of today's dollars. Since such power is ineluctably intermittent (unreliable), they would also have to buy enough batteries to keep things running during the roughly 73.6% [100-26.4] of the time that their solar panels would not be producing much. How many batteries would that be? Assuming that Africa's future inhabitants decide that they could get by with just 1 day's-worth of energy storage (that's not conservative- widespread cloudy and windless periods often last longer than one day) they would have to build/buy about 216 billion kWh's worth of storage capacity [2000 J/s*4.5E+9*3600 s/hr*24 hr/day/3.6 E6 J/kWh = 2.16E+11 kWh], which, if implemented with Tesla's equally real and state-of-the-art 13.5 kWh, ~$7000, lithium ion battery "Power Walls", would cost today's subsistence farmers' hopefully much more prosperous descendants another $100 trillion [$7000/13 kWh*200E+9 kWh/3.6E+6 J/kWh] of today's dollars to purchase the first time around (lithium ion batteries only last for a few years). Real world windmills exhibit similar capacity factors meaning that if they were to be used instead, a similar amount of energy storage capacity or some other sort of clean backup power would be required. Natural gas – today's primary backup for wind and solar power plants – will probably be prohibitively expensive by then because all of the "cheap" natural gas will have already been discovered/fracked/consumed. Leaving it in the ground along with most of the world's remaining coal and oil would be a very good idea because burning them would otherwise add to the already excessive amounts of anthropogenic greenhouse gases (GHGs) responsible for climate change – coal as CO_2 and methane both as-is[19] and after it's been oxidized to CO_2.

The above-derived numbers are actually somewhat optimistic because Africa is a better site for most today's renewable energy sources than is much of the already-developed world. For example, by 2014 European

---

[19] While carbon dioxide is typically painted as the bad boy of greenhouse gases, methane is initially about 100 times more potent as a heat-trapping gas and several percent of that currently being fracked leaks directly into the atmosphere [246]. Any gaseous substance's mass-wise Global Warming Potential (GWP) relative to CO_2 depends on the timespan over which that potential is calculated. A short half-life gas which is quickly removed from the atmosphere may initially have a large effect, but over longer time periods, become less important. Thus methane's GWP over 100 years is about 34 but 86 over the first 20 years. Agriculture's contribution to anthropogenic GHG emissions (~9%) is primarily due to methane (cow farts/burps, rice fields, etc.), not CO_2 emitted by farm machinery engines.

Union countries had invested approximately €1.1 trillion (about 1.4 $trillion) in large scale renewable energy installations – mostly wind turbines and solar panels. That money provided a nominal nameplate electrical generating capacity of about 216 Gigawatts, nominally ~22% of total current European energy demand (~1000 Gigawatts). Data supplied by Europe's Renewables Industry indicate that its total output throughout 2014 averaged 38 Gigawatts (~ 3.8% of Europe's electricity requirement) at a combined mean capacity factor of ~18%. When adjusted for that factor, the capital cost of Europe's wind/solar energy installations was €29 billion/GWe - about 30 times that of its conventional gas-fired electricity generation facilities and 3–10 times greater than GEN III nuclear power plants.

Since the output of most renewable energy plants depends upon the season, local weather conditions and what time of day it is, their contribution to the electricity grid is erratic, intermittent, and non-dispatchable. This means that they often can't contribute to supply when it's needed, thereby rendering them useless (decorative?) much of the time. On the other hand, rules mandating use/priority of renewable electricity cause major grid disruptions when their output suddenly rises because dispatchable thermal (fossil fuel and nuclear) power plants have to be closed down to a zero-efficiency "idling" state. Consequently, despite virtually no fuel costs, according to US EIA, renewable energy installations data can still cost up to 1.5 – 2.5 times as much to operate and maintain as do conventional gas fired power plants.

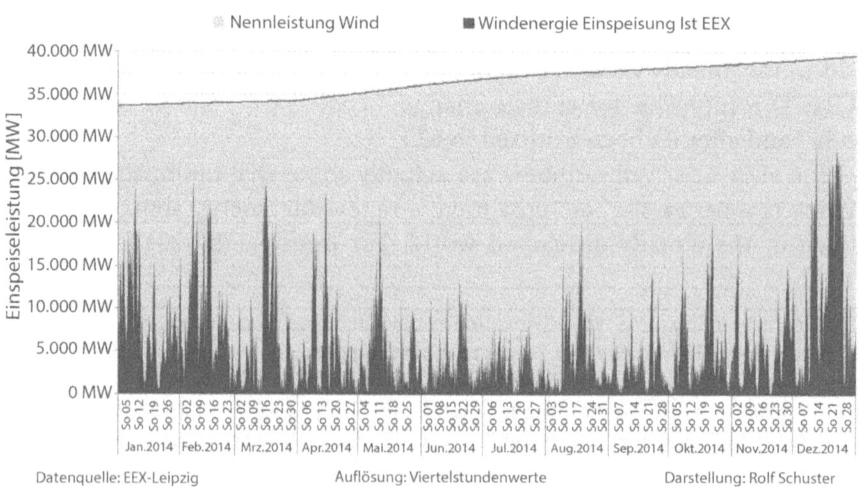

Figure 4 Germany's real time wind power generation throughout 2014 (Source: http://www.vernunftkraft.de/85-prozent-fehlzeit-windkraftanlagen-sind-faulpelze/).

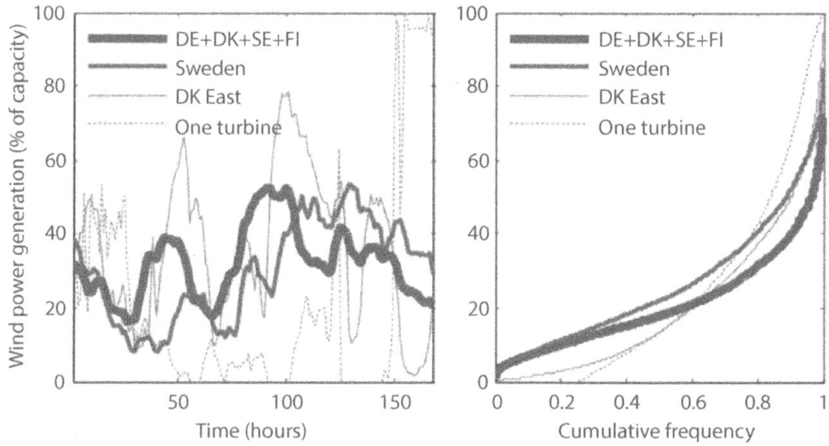

Figure 5 Typical variations in wind power generation on different geographical scales (figure courtesy of Juha Kiviluoma).

The red line across the top of Figure 4 was Germany's wind power "capacity"[20] that year – the blue spikes under it represent total power actually produced. That system's capacity factor (energy produced/nameplate capacity) was about 0.16 and there were many times when essentially no power was generated across the entire country.

The leftmost figure of Fig. 5 displays one week's worth of European wind power data during 2010. The right shows two years (2010 and 2011) of hourly data sorted by generation level. The aggregated generation for four European countries was created by calculating a weighted average of the capacity factor for each hour (Germany's weight was three because it is a bigger country; the others were assigned a weight of one).

South Africa's future plans [41] predict that by 2050 its electrical supply will be as follows: 20.4 GWe of nuclear capacity (up from today's ~1.8 GWe) is to supply 30% of its electricity from 14% of the country's total generating capacity, coal will generate 31% from 18% of total capacity, wind 18% from 37.4 GWe of capacity, and 6.5% from 17.6 GWe of solar PV capacity (the rest of its electricity, about 14%, will primarily be imported hydropower). Combining those numbers tells us that South Africa's total anticipated generating capacity by then will be about 161 GW [20.4/.14/0.9 where 0.9 approximates a LWR's CF] – over three time's today's - and the relative capacity factors for those sources will be as listed in the third column of Table 3.

---

[20] nameplate capacity is the figure usually quoted by wind & solar power salespersons & enthusiasts.

Table 3 South Africa's electrical supply circa 2050.

Source	%Energy delivered/%capacity	CF/CF nuke
Nuclear (GEN II or III LWRs)	2.143	1
Coal fired thermal	1.891	0.882353
Wind	0.481	0.224599
Solar PV	0.369	0.172348

(based upon figures from [41]).

"Unfiltered" data like those in Figs. 4 and 5 demonstrate why it is both irresponsible and callous for some of the "first world's" energy experts to continue to insist/pretend that any combination of intrinsically unreliable renewable energy sources – windmills, solar panels, etc. – could provide the energy required by 11.2 billion people possessing a fair/equal share of a totally connected, cleaned-up, and "rich" technological civilization.

While it would be possible for "rich" people living in temperate climates to be comfortable when they're within their well-insulated, properly windowed/oriented homes outfitted with especially efficient appliances and 7-20 thousand dollars' worth of "Power Walls", the technological civilization that they would dependent upon for everything else (transportation, manufactured goods, and, food) will require reliable, not intermittent, power.

2.3.1 Which Food Crops Should Our Descendants Raise and How Much Land Would That Take?

Next, we must decide what my example population's (Africa's) should eat and how much land would be required to provide that food. Since vegetarian diets are much more efficient resource-wise than are those generally consumed by today's richer people, for simplicity's sake, I'll assume that by 2100 AD everyone will be consuming 2500 kcal/day (1.05E+7 J), most of which is provided by two especially productive crops raised upon the minimum amount of soil capable of providing yields currently achieved in the USA.

A recent paper [42] discussing the amount of land (m^2) to produce a gram of protein with different crops, USDA reports (http://usda.mannlib.cornell.edu/...) of US crop yields/acre, and candidate food crop characteristics from various WIKIPEDIA entries suggests that an efficient combination would comprise maize (corn) because it's exceptionally productive, nutritious, and already widely produced/consumed/accepted in Africa plus some sort of pulse (legume) to complement its unbalanced mix of amino acids (not enough lysine - [43]). Of the likely pulses, peanuts seem to make the most sense because they are a "hot weather crop", taste considerably better, contain more fat/oil, also already widely produced/consumed/accepted in Africa, and almost equally important, would not extract as much phosphorous and potassium (key macronutrients) from its soil per food-calorie as would the next runner-up crop, soybeans.

Assuming zero waste, providing 2500 kcal/day of food for 4.5 billion people translates to 4.1E+15 kcal (1.72E+19 Joule (J)) worth of foodstuffs per year. If we also assume that 75% of their food-type calories are to be provided by maize, simple calculations indicate that the total amount of land required to feed every African person circa 2100 AD adds up to 9.71E+7 hectare(ha), of which 7.28E+7 ha would be devoted to maize and 2.43E+7 ha to peanuts.

That combination would provide everyone with 59 grams of "complete" protein per day along with virtually everything else that humans need to first grow up and then remain healthy[21]. Africa's folks would probably also want to (and should) devote perhaps an additional 5-10% of similarly productive/managed land to raising the lower calorie/protein but tastier fruits, vegetables, and spices that render vegetarian diets far more palatable than most of the world's "rich" people realize. Additionally, if Africa's much more prosperous (on the average) future inhabitants were to decide that chicken should provide 20% of their food calories (500 kcal/day/person – about fourteen times more than they currently consume), similar calculations suggest that roughly 10% additional land would be required to raise the peanuts and maize needed to feed those birds as well. The substitution of the purportedly equally nutritious/delicious cricket "meat" [44] for that chicken would require only about 5% more peanuts/corn/land than would a strictly vegetarian dietary.

9.71E+7 ha is about the same amount of land that the USA currently devotes to producing all of the crops listed in Clark and Tillman's paper to support ~320 million people (~7% of the number assumed herein for Africa circa 2100 AD). Thanks to artificial fertilizers, improved crop

[21] Of course I'm just doing semi-quantitative theorizing here because people typically waste about one third of their food which means that my calculations similarly underestimate the amounts of land, water, fertilizer, etc. needed to feed their descendants. It's probably also unreasonable to assume that the world's agricultural entrepreneurs would ever permit their elected representatives to eliminate bioalcohol, palm oil, and biodiesel subsidies.

genetics, and pesticides, today's modern farmers now need 68% less land to produce a given quantity of food than did their mid 20th century predecessors utilizing that era's more "natural" (aka more "organic") farming practices. A team led by Professor Andrew Balmford of Cambridge University recently showed that the best way for humanity to preserve the world's biodiversity is to minimize the amount of land used to serve its own needs and thereby allow the set aside ("sparing") of large areas capable of supporting natural lifeforms; i.e., the set aside of arable ("good") land, not just additional desert [45]. Heading off today's ongoing "sixth extinction" will require efficient, energy intensive, farming, not switching back to inefficient natural/organic farming.

A FAO estimate [46] of the area under "*managed water and land development in Africa totals some 12.6 million ha, equivalent to only 8 percent of its arable land*". Since Africa's total area is about 30.3 million km^2, this suggests that 1.58E+8 ha [12.6e6 km^2/100 ha/km^2/0.08 or 5.2%] of it is considered to be "arable". Since my estimate of the total area required to feed its 4.5 billion future inhabitants (9.71E+7 ha) represents only 61% of that figure, hopefully, they will choose to continue to share some of their continent's still-useful land with its iconic suite of wild animals.

2.3.2 The Whys and Costs of Desalination

Like wind and solar power, water is a renewable resource with both highly variable and limited availability. Rainfall, temperature, evaporation, and runoff determine its total availability and human decisions determine who gets what. Nearly every country in the world experiences water shortages during part of the year, and more than 80 of them suffer from serious water shortages. Clean water resources per capita are declining rapidly as human population increases. Population growth reduces water availability per person and stresses the entire environmental system. Pollution, erosion, runoff, and salinization associated with irrigation, plus the overall inefficient use of water, contribute to the decline in water resources. Allocation of increasingly scarce fresh water generates conflicts between and within countries, industries, and individual communities with the majority of it everywhere consumed by agriculture. Water shortages are also severely reducing biodiversity in both aquatic and terrestrial ecosystems [47].

Africa's 4.5 billion(?) future inhabitants wouldn't be able to feed themselves with ~60% (or all) of its arable land unless they become able to irrigate it. Because irrigated land almost always produces higher yields than do rain fed farms and permits double and sometimes even triple cropping in warmer regions, such lands currently provide around 40% percent of global cereal supply [285]. Only ~4% percent of Africa's cropland is currently irrigated due to prohibitive costs, insufficient water, and a general lack of commitment to infrastructure-related investment in things like power plants and the fuel needed to operate them. Consequently, it's unlikely that this book's cornucopian future would be implemented by either the African people themselves or the institutions/people that have provided most of their "aid" to date[22]. However, let's pretend that will happen.

Pumping water onto approximately 10 percent of the world's total arable land (around 300 Mha) currently consumes around 0.225 EJ/yr. Another 0.05 EJ/yr of indirect energy is devoted to the manufacture and delivery of irrigation equipment [48]. Around two-thirds of the water currently used for irrigation is drawn from underground aquifers. Energy intensive electricity powered deep well pumping accounts for about two-thirds of that and projections suggest that it will become ~90% by 2050 when shallow reserves are almost totally depleted. Current aquifer water extraction rates exceed recharge rates – grossly so in many places. Additionally, global warming is simultaneously exacerbating droughts and melting the glaciers that feed the rivers providing much of the world's cheap-to-deliver irrigation water. Global warming has caused Mount Kilimanjaro's "snows" to disappear along with most of those within the USA's Glacier National Park. Building more dams won't solve such problems because dams do not create water. Additionally, a comprehensive review of Nigeria's outside-funded dam projects [49] concluded that while they do make money for local promoters and the outsiders that fund/support them, they decrease net agricultural productivity by turning fertile downstream floodplains into deserts. In addition to killing wetland-dependent wildlife[23], those dams have served to lower not raise, the incomes of far more people than have benefitted.

[22] It really just boils down to human nature. Rich people usually have lots of alternatives and choose the cheapest approach to solving their problem whereas poor people without such options have to live with theirs even if doing so will eventually kill them (who cares? - most poor people live" *in ...thole countries*" and aren't the right color anyway). Building enough desalination facilities and their power plants to "save" the world's poor people will cost several hundred $billion & it's pretty darn unlikely that today's venture capitalist dominated societies will ever do anything that doesn't guarantee them and theirs a "reasonable" return on investment.

[23] Dams have almost totally destroyed salmon runs throughout most of the "civilized" world including, of course, my home state of Idaho. Its trout streams have been further impacted by a myriad of tiny dams and extra diversions built to take advantage of the fact that low head hydropower plants are heavily subsidized. Other things that have served to devastate its trout fishing include the now almost universal use of neonicotinoid pesticides [269] and refusal to enforce a law requiring irrigation canal companies to screen their headgates [247].

Most such dams also don't generate nearly as much electrical power as "promised" due to inadequate maintenance and low (water-limited) capacity factors. Finally, at best, dams represent a temporary fix for the problems that they are built to address because their reservoirs will eventually fill with mud.

Water shortages plus the high cost of desalination – primarily due to high electricity costs – has led some countries rich enough to do so (e.g., China) to reduce their own crop production and rely more heavily upon imported grains.

This situation is unsustainable which means that I'll next assume that most of the water irrigating Africa's future farmlands (and much of the rest of the world's too) would be generated by desalinating seawater – the Earth's only truly inexhaustible/sustainable water source. Wikipedia's description of Israel's solution to its water issues [50], demonstrates how a properly managed future technological society could address the whole world's water-related woes. Israel's ~8.5 million people are fed by ~1.045E+9 m^3 of fresh water applied to its mostly-irrigated farmland [51]. This suggests that the rest of the even more water-stressed Middle East's mostly-poor ~101 million people, including the majority of those living in Palestine, Iraq, Jordan, Lebanon, Oman, Syria, and Yemen [52], could be equally well supported by irrigating their potentially arable land (assuming all of it) with 1.24E+10 m^3 of desalinated seawater. Assuming the ~3kWh/m^3 energy requirement [53] of today's most popular approach to desalination, reverse osmosis (RO), doing so would require an energy input of 1.34E+17 joules/a, which corresponds to the full-time output of ~4.2 one-GW_e nuclear reactors. The volume of water corresponding to adding 0.51 meter (20") of it over 9.71E+7 ha of African farmland is 4.93E+11 m^3, which, if generated via RO, would require the full-time output of ~169 full-sized nuclear reactors (5.33E+18 Je/a). In principle at least, Siemen's electro dialysis-based desalination technology would require only about one-half that number of reactors and is also less apt to become fouled by seawater's other-than-salt impurities [54].

To continue, Africa's average elevation is about 600 m (2,000 feet) above sea level, roughly the same as that of both North and South America. Assuming that all of Africa's desalinated irrigation water would have to be pumped uphill that far, the energy needed to do so would be 2.90E+18 joules [4.97E+11m^3*1000 kg/m^3* 600 m*9.8 m/s] requiring another 92 full-sized power plants.

How much would Africa's desalination equipment cost? The contractual cost of the world's (Saudi Arabia's) biggest (~one million m^3/day), RO-based desalination plant is $1.89 billion [55]. Collectively, these numbers suggest that building enough RO plants to irrigate Africa's future farmlands would require a one-time capital expenditure of $2.55 trillion [4.93E+11*0.00189/(365*1E+6] – about 11% of the USA's current national debt. Similarly addressing California's Central Valley's chronic irrigation water problems should cost only about $40 billion.

As mentioned earlier, the Western World's recent Middle East military incursions will probably end up costing its citizens ~thirty times more ($4–6 trillion) than it would to have built enough nuclear powered desalination plants to provide sufficient fresh water for everyone living there and thereby address a root cause of that region's almost perpetual turmoil. For example, a recent paper in the Proceedings of the (US) National Academy of Sciences [261] points out that the chief driver for today's Syrian conflict is the unrest/poverty generated by relentlessly worsening droughts, not a desire for "regime change".

Another plus for desalination is that its product does not add additional salts to soil and is also better at remediating already over-salinized soils than is ground water. A final plus is that because it doesn't already contain near-equilibrium levels of calcium, magnesium, carbonate/bicarbonate, silica, etc., it is a better rock solvent (more corrosive) than is ground water. The next section will reveal why this is important.

2.3.3 Fertilizers

Another reason that the productivity of Africa's farmlands is considerably lower than that of more developed regions is that relatively little fertilizer is used. The three most important components of fertilizers (macronutrients) include nitrogen in either its negative three (ammonia-type) or positive five (nitrate-type) oxidation states, potassium (invariably in its plus one oxidation state), and phosphorous (invariably in its plus five oxidation state). Nitrogen fertilizer production currently accounts for about one half of the fossil fuel (mostly natural gas) used in primary food production.

2.3.3.1 Nitrogen and the Cost of Fixing Enough of it to Feed Everyone

We'll start with nitrogen. US farmers hoping to produce 13.1 t/ha (200 bu/acre) of maize (corn grain) are advised to add ~ 258 kg of N/ha [56]. Since peanuts (a legume), can recover/fix its own nitrogen from air, much less nitrogenous fertilizer would be needed for land devoted to its cultivation – let's say 50 kg N/ha. Assuming those application rates, fertilizing Africa's future crop land would require ~ 2.0E+7 tonnes of ammonia each year (one kg N ≈ 1.21 kg of ammonia) A recent review of synthesis technologies [57] concludes that each tonne of ammonia made with electrochemically-generated hydrogen, pressure swing-generated atmospheric nitrogen, and conventional Haber Bosch processing equipment would require about 14.2 MWh's [14.2*3.6E+9 J] worth of electricity. That, in turn, suggests that satisfying this scenario's nitrogenous fertilizer requirement would require the full-time output of ~32 one-GWe power plants.

2.3.3.2 The Reasons Why Powdered Basalt Should Supply the Phosphorous and Potassium Required to Feed Everyone

Since...

- Much of Africa's (and the rest of the world's) farmland has already lost a great deal of its topsoil via erosion,
- Much of its remaining topsoil is mineral-depleted,
- Basic (mafic) rock-weathering is how Mother Nature limits the Earth's atmospheric CO_2 concentration (figure 6) via "mineralization",
- Basaltic rocks are both intrinsically basic (contain a good deal of magnesium and calcium) and rapidly weathered by the natural phenomena extant in cultivated soils [58],
- Most of the Earth's crust consists of basaltic rock, a good deal of which is either on or close to the surface of its continents,
- Basaltic rocks contain relatively high concentrations of potassium and phosphorous along with all of the other biologically important elements which explains why soils comprised primarily of weathered volcanic ash most of which originally consisted of molten basalt are exceptionally productive [59],

...we'll next assume that the phosphorous and potassium required to produce Africa's food crops circa 2100 AD will be provided by amending its farmland with powdered basalt. In order to be effective, any such amendment must weather rapidly enough to release sufficient potassium and phosphorous to support high-yield agriculture which, in turn, means that the raw silicate rock surfaces must be "fresh" (not already equilibrated with the atmosphere and thereby covered/blocked with secondary phases), ground to a considerably smaller particle size than is the quarry waste-type soil amendment rock powder currently being marketed to hobby farmers, and also mixed with root-zone topsoil, not just dumped upon the surface of

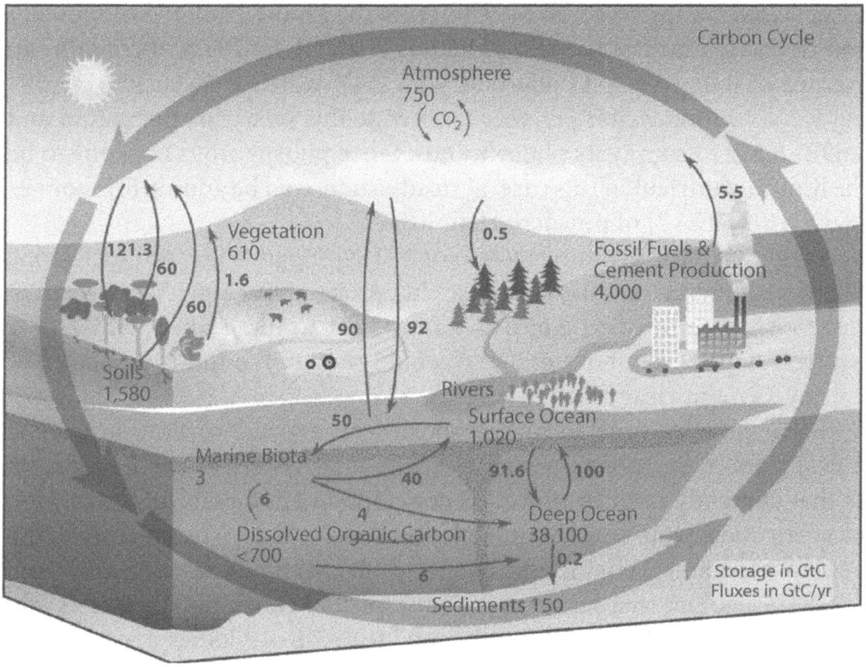

Figure 6 The Earth's carbon cycle (commons Wikipedia.com).

the ground [60, 61]. Based upon the rather limited amount of scientifically planned/supervised experimentation described in the open-access (not paywalled) technical literature, I'm next going to assume that this would require grinding it so that the particles comprising >80% of it possess diameters <10 microns. Since rock grinding is highly energy intensive – much more so than is simply recovering it from a quarry's rock outcrop or waste dump – the cost estimate for this part of my scenario will be based upon that step's energy demand plus the resulting powder's transport and distribution costs.

First, how much powdered rock must be made? The food stuff P and K concentrations, land areas, and crop yield figures in the papers referenced earlier suggest that the food consumed each year by Africa's 4.5 billion future inhabitants would contain 2.64E+9 kg of potassium and 2.05E+9 kg of phosphorous. Assuming (wrongly I hope, but consistent with the way that things usually happen) that neither nutrient is subsequently recycled back to the soil, both must be replaced each year via basalt weathering. The compositions of flood basalts vary considerably but since all originate from the Earth's fairly well mixed underlying magma, for the following estimates I'm going to assume a composition with which I'm familiar [62, 63] – that of the basalt comprising Idaho's "Craters of the Moon" National Monument and covering much of the rest of Idaho's Snake River Plain. It contains an average of 0.61 wt% K_2O and 0.55 wt% P_2O_5 which translates to requiring 5.21E+8 tonnes of it per year to provide this scenario's potassium and 8.69E+8 Mg to supply its phosphorous. Since phosphorous happens to be the limiting nutrient in this case, at steady-state, we'd be adding 8.95 tonnes [8.69E+8/9.71E+7] of powdered basalt/ha/a.

A review of rock grinding technologies [64] suggests that producing one Mg of <10 micron basalt powder would require about 100 kWh's worth of electricity. If so, making 8.69E+8 tonnes of it would require 3.13E+17 J, which if done throughout one year would require the full-time output of 9.9 one-GWe nuclear reactors.

If that powder were to be transported an average of 1930 km (1200 miles) from mine-to-farm via an electrified rail system as energy-efficient as that currently used to move US coal (185 km/L diesel fuel/short ton), its energy cost would be about 6.80+16 J (assumes 1.1 Mg/short ton, 33% heat-to-mechanical engine efficiency and 44.5 MJ/kg diesel fuel with a SpG of 0.85). Trebling that figure to account for fuel consumed by trucks and tractors at the rail heads, brings the total to 2.04E+17 joules/a, which corresponds to an annual transportation/distribution energy demand requiring another 6.5 one-GWe nuclear reactors to satisfy.

An application rate of 8.95 tonnes/ha/a is not really very large because it represents only about 0.5% of the mineral matter already within a six inch deep (typical annual crop root zone) layer of normal density/composition topsoil and is also considerably less than what current farming practices often lose via wind/water erosion (typically about 30 t/ha/a - Pimental 2009). Consequently, since my scenario's rock grinding/distribution costs are much lower than its irrigation water and nitrogenous fertilizer costs, it would be a good idea to at least start out with considerably larger powder application rates, perhaps 40-50 tonnes/ha. Doing so would also reduce the chance of crop "starving" due to slower-than- assumed weathering rates.

The 309 [169+92+32+9.9+6.5] GW_e's worth of "clean" power plants required to implement the agricultural aspects of the above-described African scenario's clean/green utopian future is about the same amount of power currently generated by all of the world's nuclear reactors and about four times greater than Africa currently produces in any fashion (about 650 TWh, see [65]). However, it represents only ~3% of the total energy services required by 4.5 billion people as rich as the EU's are now.

All of the necessarily huge machinery and manufacturing facilities required to implement this or any other technological fix capable of "saving the world" would be much cheaper to build and operate much more efficiently with reliable power than with that provided by intermittent sources. While it would indeed be "possible" to run desalination/ammonia plants, rock crushers, tractors, locomotives, etc., with windmills and/or solar panels, doing so would be expensive, dangerous, and frustratingly unproductive to such machinery's owner-operators and their customers. It would also require that 1/CF times as much machinery be built to do the job at the same rate that a CF ≈ 1.0, MSR-powered world could: typically ~3 times as much machinery for wind and 4–10x as much for solar-sourced power.

Again, intermittent power supplies are suitable for some niche applications (e.g., charging a terrorist's cell phone), not for powering technological civilizations [66]. That's why today's farm tractors, locomotives, container ships, cruise liners, and air liners are fossil fueled, not wind powered.

Also again, the above ball park numbers are just approximations because the rate and degree to which powdered basalt would release its constituents (weather) under field conditions is affected by a host of factors/variables. A nutrient-specific discussion of some of them may be found in a FAO report describing the use of raw phosphate rock as fertilizer [67]. Thankfully, that subject is beginning to receive a good deal of attention [68] and some hopefully realistic experimental studies have begun [59].

3

A Sustainable Nuclear Renaissance's Other "Killer Apps"

3.1 Atmospheric Carbon Sequestration

Let's begin this section with another especially revealing ball park calculation:

Wikipedia entries tell us that climate experts have concluded that the effect of doubling the atmosphere's CO_2 concentration corresponds to a forcing factor of ~3.7 Watts/m^2 at the earth's surface.

There's also a consensus that mankind has dumped about 580Gt of carbon into the atmosphere since the beginning of the industrial age and that it's apt to cause a mean global temperature rise of from 1.5 to 4.5 Centigrade degrees – for the purpose of this example, let's say 3 degrees/doubling.

If we assume that that carbon was all derived from burning petroleum with composition $(CH_2)x$, that's about 677 Gt [580*14/12] tonnes of it.

At 42 MJ/kg, the heat energy we got by burning it amounted to ~2.84E+22 Joules [677*1E+12*1E+3*42E+6].

Today's ~410 ppm atmospheric CO_2 concentration corresponds to about 46% of a doubling of the world's pre-industrial, ~280 ppm level.

Since the Earth's surface area is about 5.2E+14 m^2, that corresponds to an anthropogenic carbon heat input rate of 8.85E+14 W [0.46*3.7*5.2E+14].

The point of this little exercise was to show that any such fuel's GHG-warming effect's "bad" energy production will exceed its useful combustion-generated heat energy within ~0.98 years [8.85E+14J*3.15E+7 s/a/2.84E+22J] and is likely to continue to do so for at least another 50 years (the "half life" of CO_2 in the atmosphere).

Energy generated building/using nuclear power plants results in the release of relatively little GHG/J – considerably less than that generated with wind turbines or solar panels (Fig. 4) – and the future's more thermally efficient and more compact reactors would generate even less.

Darryl Siemer (ed.) Nuclear Power: Policies, Practices, and the Future, (35–50)

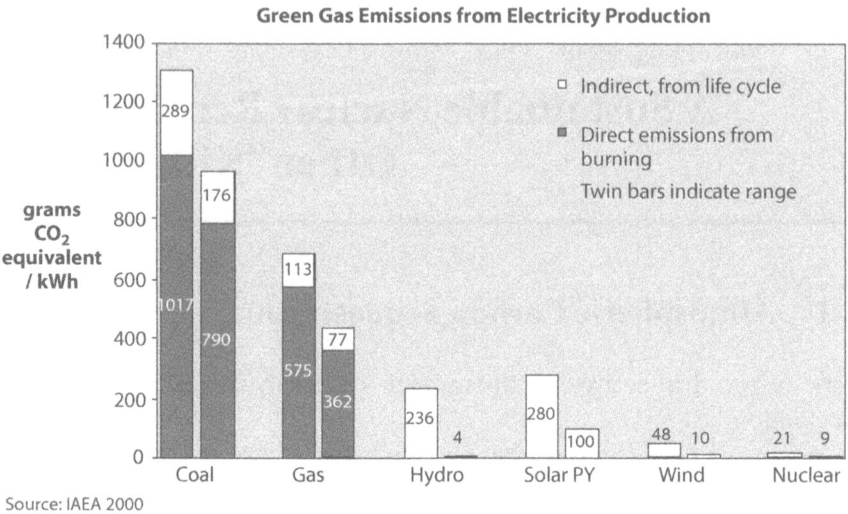

Figure 7 Greenhouse gas emissions of various energy sources (a more up-to-date but too "fuzzy" to reproduce such figure along with its literature citations may be seen on p. 35 of, https://www-pub.iaea.org/MTCD/Publications/PDF/CCNAP-2018_web.pdf).

The speediest drop in greenhouse gas pollution on record occurred in France in the 1970s and '80s, when it transitioned from burning fossil fuels to nuclear fission for electricity production which lowered its greenhouse emissions by roughly 2 percent per year.

According to James Hansen *et al.* [71] the task facing the entire world today is more difficult because *"emissions reduction of 6%/year plus 100 GtC storage in the biosphere and soils are needed to get CO_2 back to 350 ppm, the approximate requirement for restoring the planet's energy balance and stabilizing climate this century"*. His colleague & co author, Jeffrey Sachs, director of the Earth Institute at Columbia University says that, "*On a global scale, it's hard to see how we could conceivably accomplish this without nuclear*".

It been known for almost two centuries that the ultimate sink for the atmosphere's carbon dioxide is basaltic (basic) crustal rock which, when exposed to atmospheric CO_2 and moisture, eventually weathers to form the miscellaneous oxides, clays, feldspathoids, and carbonate minerals that make up much of the inorganic components of soils [256]. Soil holds more than three times as much carbon as the atmosphere [72], yet its potential for deliberately reducing atmospheric carbon-dioxide levels (CDR) and therby mitigating global warming, although much studied, is not being exercised [59]. When completely weathered by those mechanisms collectively responsible for it in biologically active soils, each gram of my example's basis basalt (10.06 wt% CaO and 7.65 wt% MgO) would release 7.35 milliequivalents [0.1006*2/(40+16)+0.0765*2/(24.32+16)] worth of base. If we assume that such base would convert acidic soil-gas CO_2 that would otherwise transpire (to the atmosphere[24]) to the bicarbonate anion [73], the application/weathering of 8.95 t/ha of such basalt over 9.71E+7 ha of African farmland would remove/sequester 0.076 Pg (76 million tonnes) of carbon. That sounds like a lot of "sequestration" but represents only ~0.009% of that currently in the atmosphere (~3300 Gt CO_2).

If carbon sequestration is to become a primary goal, another way to go about doing it would be convert corn stover[25] and peanut hulls/stems (their leaves would probably end up on the ground) to "biochar". Assuming this scenario's African grain crops, that translates to converting about ~11.6 tonnes of biomass to ~3.1 tonnes of biochar and 5 tonnes of bio oil per ha [74, 75]. Because biochar is ~70% elemental carbon, burying it would simultaneously increase Africa's soil's organic matter (SOM) and sequester atmospheric carbon at the rate of ~0.25 Pg (250 million tonnes) per year.

If everyone in the world circa 2100AD –not just Africans - were to char their stover and fertilize with Snake River Plain basalt at the rate determined above, they would collectively sequester the equivalent of about 3 Gt CO_2 per year. However, since the atmosphere already contains about 500 Gt of excess CO_2 [(~412ppm-350ppm)/412ppm*3300 Gt =496] and will surely get further unbalanced before we kick our addiction to fossil fuels, it would probably take over two centuries for the future's farmers alone to reduce it to a "safe" (350 ppm) level. This of course, suggests that along with stopping atmospheric pollution and switching to the Earth's most natural fertilizer, a "nuclear clean new deal" must accomplish something else as well.

What would that be?

The Earth's oceans have become an especially tragic "commons" with respect to the impact of CO_2 dumping [76] via acidification and oxygen loss due to water-warming. This is killing a host of pelagic creatures with aragonite (calcium carbonate) skeletons/shells all of which depend upon oceanic

[24] This phenomenon is termed "respiration". It does not necessarily represent a net transfer of carbon from soil to atmosphere because it is largely offset by vegetative carbon inputs to soils. It is biologically driven (plants suck CO_2 down into the soil) and currently releases about eight times as much CO_2 to the atmosphere as does our fuel burning [251]. However, it is soon apt to be releasing much more because global warming is upsetting its normal equilibrium especially in the Arctic. Many scientists believe that a catastrophic "tipping point" is imminent.

[25] Stover is above-ground, non-grain, crop matter - there's generally about as much of it as grain and approximately 80% of it can be readily collected.

chemistry (pH and temperature) remaining as it has been for ~50 million years. These calcifying creatures constitute the Earth's dominant natural CO_2 sequestration mechanism, converting ~1 billion tons of CO_2 each year to oceanic sediments[26] and limestone (coral reefs). Land plants and soils don't come close because when today's farmed soils aren't amended with something like powdered basalt, soil microorganisms actually add (via respiration) net GHG to the atmosphere via their metabolism of soil organic carbon derived from the plants grown in them. Modern civilization's sudden conversion of the Earth's fossilized carbon to atmospheric CO_2 is driving oceanic extinctions that will eventually eliminate species ranging from coccoliths to whales, and thus about 15% of all human food protein.

A relatively inexpensive and practical way to address those effects would be implement the suggestion proffered by Professor Schuiling and his colleagues well over a decade ago (e.g., [78]): i.e., crush basalt, or maybe even better, dunite (it's more basic[27]), into coarse sand-sized particles and scatter it/them along coast lines and shallow reefs. When so situated, natural wave action would greatly accelerate such sand's weathering and thereby immediately relieve over-acidification while rendering the oceans better sinks for CO_2 and thereby affecting its removal from the atmosphere as well.

Another and maybe more effective driver for the implementation of Drs. Schuiling & Krijgsman's proposal is that a global sand shortage has come to be because construction industries use it to make brick, asphalt and concrete – especially in and around Southeast Asia's burgeoning urban areas. Consequently, vast amounts of it is being "stolen" from beaches and river banks at the same time that rising sea levels and climate change-induced river flooding is increasingly threatening homes and businesses situated in the denuded areas. That's apt to render artificial basalt/dunite sand valuable enough to tempt entrepreneurs to make/sell lots of it when power becomes cheap enough.

Saving especially valuable waterfront-situated homes & hotels is apt to be at least as strong a motivator to important people and their elected representatives as is making everyone equally energy "rich", rendering agriculture sustainable, and/or protecting any sort of commons.

---

[26] About 80% of the carbon stored in the geosphere is limestone and its derivatives formed from the sedimentation of calcium carbonate stored in the shells of marine organisms. The remaining 20% is stored as kerogens/oil/gas etc. formed through the sedimentation and burial of terrestrial organisms under high heat and pressure [77].

[27] Dunite is an ultramafic plutonic rock possessing a chemical composition (majors only) falling somewhere between pure forsterite (Mg_2SiO_4) and pure fayalite (Fe_2SiO_4). High magnesium dunite is about four times as basic as typical flood basalt. It's a better CO_2 absorber but a poorer fertilizer.

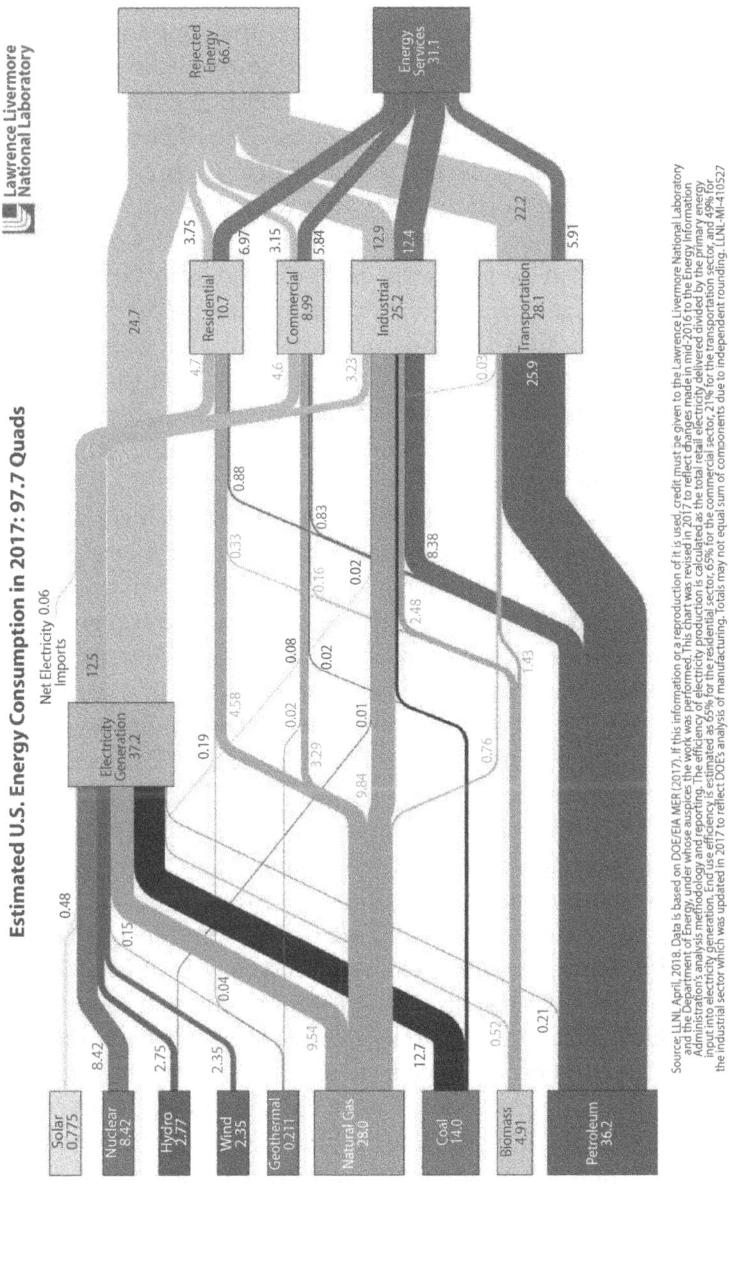

Figure 8 LLNL's 2017 US energy Sanky diagram.

3.2 Nuclear Powered Transportation

3.2.1 Direct Electrical

LLNL's 2017 energy flowchart (above) indicates that amount of fossil fuel consumed by the USA's non nuclear thermal power plants (22.45 quads worth [9.54+12.7+0.21]), is about 85% of that consumed by its transportation system ~100% of which is petroleum-based. The efficiency with which its electrical power plants convert their fuels' heat energy to electricity is about 40% while its transportation system is only about 21% [5.91/28.10] efficient. Consequently, ignoring other losses replacing the USA's fossil fuel powered transportation system with one powered with electricity generated by 40% efficient thermal-to-electric nuclear power plants would require 7.76 [5.91*21/40/0.4] quads of such heat. That's almost exactly what's currently generated by its ~100 reactor LWR fleet (8.42 heat-type quads). At 3.2E-11J/fission, doing this would require the fissioning of 96 tonnes of uranium per annum (a 5.5 ft cube if it's metallic) which would generate 96 tonnes of fission products.

Electricity represents an almost ideal future transport fuel [79] – lightweight electrified vehicles for local passenger and freight moving and high speed trains for almost everything else. Unlike other alternative transport energy scenarios, only electric mobility can move people and goods using any combination of raw energy sources –hydroelectric, wind turbines, and photovoltaic panels or gas turbines powered with coal, natural gas, oil, wood waste, solar energy, bio oil or, preferably, nuclear fission. Energy-wise, the majority of these vehicles would be grid-connected (GCVs) meaning that their electricity is generated remotely and delivered directly by wire or rail to its motor(s). GCVs currently do the majority of electrified people/freight moving. Electric streetcars and trains were operating in many cities by the end of the nineteenth century and ~150 cities around the world have or are developing electric heavy-rail (e.g., metro and commuter rail) systems running at either the surface, elevated, or underground. Some 550 cities in Europe and Asia have streetcar and/or light-rail systems and about 350 have trolley buses. Electrification of intercity railroads began early in the twentieth century, although most of it occurred after 1950. Now most rail routes in Japan and Europe are electrified. Russia has the most extensive system; approximately half of its 85,000 kilometer total, including the whole of the 9,258-kilometer Trans-Siberian Railway is electrified. China's rail system is being rapidly electrified and now boasts the world's second-most extensively electrified transport system: 49 lines totaling about 24,000 kilometers. In these countries and elsewhere, those are mostly main routes and thus carry a disproportionately large share of their county's passengers and freight. The revolution caused by introducing high-speed electrified passenger rail has transformed the way that people move between major cities in Japan and Western Europe. Their primary advantage relative to battery-electric vehicles (BEVs) is lower cost and greater efficiency. No matter how good any powerful/long range BEV might become, it would still have to carry around a huge load of very expensive batteries which will take up space and increase its energy consumption. For example, a 85 kWh, "TESLA 3" car battery weighing 478 kg costs about $30,000 and stores as much primary energy as does about 2 gallons of diesel fuel. AGCVs would either need no batteries or only relatively small ones for limited "off-wire" travel. A GCV is subject only to energy distribution losses in moving the electricity from its source to the motor. For a BEV, losses incurred during the charging and discharging of its battery would likely be several times that distribution loss.

Nevertheless, there would definitely be a strong demand for small lightweight BEVs suitable for short range commuting, grocery shopping, etc. They should be designed around a government-standardized 10-20 kWh battery which could quickly switched at "filling stations" as envisioned by Thomas Edison over a century ago.

Other types of GCVs have been and continue to be used to move goods. These vehicles include diesel trucks with trolley assist such as were used in the Quebec Cartier iron ore mine from 1970 until that mine was worked out in 1977. These trucks were in effect hybrid vehicles with electric motors powered from overhead wires providing additional traction when heavy loads were carried up steep slopes. A diesel generator provided their electricity. The result was an 87 percent decrease in total diesel fuel consumption and a 23 percent increase in productivity.

Several direct comparisons of raw/primary energy consumption by GCVs with similarly capable vehicles with diesel-engine drives confirmed that energy use at the vehicle is invariably lower. For example, in 2008 San Francisco electric trolleybuses used an average of 0.72 megajoule of energy per passenger-kilometer; in contrast, the average for diesel buses in the same city was 2.67 megajoules per passenger-kilometer.

If the electricity powering trolleybuses were to be produced by a diesel generator operating at 35 percent efficiency, with 10 percent distribution loss, the buses would still use less energy overall than do conventional direct diesel-powered buses. When electricity is produced renewably (e.g., by my thought experiment's sustainable nuclear fuel cycle) the only thing that would count is such vehicles' energy demand.

3.2.2 Nuclear Powered Transportation Fuel Synthesis

Much of what people in today's richer countries have come to take for granted depends upon a world-wide transportation system that will suffer from price shocks and shortages as petroleum production inevitably declines. Petroleum still provides about 40% of the world's total primary energy supply, the largest share. Changes in its price and availability will have tremendous impact worldwide because alternative sources don't contribute much to the world's transportation sector. Petroleum production will decline due to the same natural laws underlying everything. It is often claimed that Hubbert's "peak oil" concept – the fact that oil production via any means from any/all sources will reach a maximum level then decline, – is only about geology. In fact, it is a consequence of geology, reservoir physics, economics, government policies and politics. Their intrinsic limitations will eventually affect all human activities because neither economic incentives nor political will can break or even bend them. Several natural depletion mechanisms affect petroleum production. Depletion-driven decline occurs during the primary recovery phase when decreasing reservoir pressure leads to reduced flow rates. This phenomenon is especially prevalent in fracked-type oil and gas wells. The secondary recovery phase involves water injection to maintain pressure but increasingly more water and less oil is recovered over time. Additional invested capital and technologies, e.g., CO_2 injection, can enhance oil recovery in a tertiary recovery phase but it comes at a still higher cost. It's like squeezing water out of a soaked sponge – easy to begin with but increasing effort is thereafter required for diminishing returns. Eventually, squeezing a sponge or oil basin harder isn't worth the cost/effort and will inevitably cease.

Another natural analogy is the relationship of predator to prey populations: "easy" oil leads to increasing profits and therefore further investment in extraction capacity (easy mice means more kittens). The easiest (typically the largest) resource reservoirs are inexorably depleted – slowly with "conventional" oil wells, quickly with fracked ones. Extraction costs in terms of both energy and monetary inputs rise as production moves to lower quality deposits. Eventually, investments can't keep pace with rising costs, declining production from mature fields cannot be overcome and total production begins to fall. Additionally, regardless of capital availability or increasingly high prices, at some point, an oil well can no longer deliver net energy (most of the cats along with their kittens die). In 1982, U.S. petroleum geologist M. King Hubbert said: "*There is a different and more fundamental cost independent of the monetary price: if oil is used as a source of energy, when the energy cost of recovering it exceeds its energy content, production will cease no matter what the monetary price may be."*

Though substitute transport fuel production such as coal or gas to-liquids, will likely increase over the next several decades, it's unlikely to compensate for the decline of oil production and certainly would not address the environmental impacts of fossil fuel burning. Today's alternative energy sources are not replacements for oil. They (wind, solar, geothermal, etc.) produce electricity — not liquid fuels.

Another reason for "biocharing" the future's nuclear-assisted agricultural residues would be that doing so should simultaneously produce more than enough carbon-neutral "oil" to fuel the machinery required to run the farms and thereby become a profitable side line for their owners. Figures in a recent report about Nebraska's farm fuel costs suggest that high-input corn farming currently requires about 70 US gallons of diesel fuel/ha/a [80]. Five tonnes of bio oil purportedly has the energy content of ~37% that much No. 2 diesel oil (i.e., 2176 liters, 575 US gallons, or 82 GJ's worth) and it should be possible to convert it to a good diesel-type engine fuel [81]. Consequently, such farms would generate about seven times as much motor fuel as they consume.

A host of other transport fuel-related possibilities are raised by the fact that a properly implemented nuclear renaissance would render electrolytic hydrogen much cheaper than it is now. For instance, hydrogenation of the ~11.6 t/ha of stover that I've mentioned above would produce about 3x as much synthetic fuel oil (~4.9E+8 tonnes/a) as would making it from biocharing's bio oil intermediate [82]. Such fuel would also be carbon-neutral because it was derived from plant matter that had received its carbon from the atmosphere.

Similarly, if Africa's 4.5 billion future citizens were to produce/consume as much Portland-type cement per capita as we do now, Fisher Tropsch hydrogenation of the CO_2 so generated would produce about 8.2E+8 Mg of transportation fuel/a, which figure divided by 4.5 billion represents ~28% of current world per capita petroleum consumption rate. Unfortunately, even if that cement were to be made with nuclear powered kilns, such fuel would not be carbon neutral because limestone would still have to be calcined.

Another possibility which would not serve to dump such carbon into the atmosphere would be to make additional "nuclear ammonia" and use it to fuel engines and/or fuel cells [83]. For example, the shipping industry is beginning to evaluate ammonia as a potential carbon-free alternative to the heavy fuel oil ("bunker fuel") used in maritime transport.

To put proposals like these into proper perspective, we need to do more ball park calculations. According to the EIA, of the 7.3 billion barrels of petroleum consumed by the USA during 2017, 47% was motor gasoline, 20% was distillate fuel (heating oil and diesel), and 8% was aviation fuel – other words about 75% of petroleum is used to fuel some sort of engine. That year, the entire world consumed about 83 million barrels of it per day or 30.3 billion barrels total. Oil's raw combustion heat energy is about 6.1 GJ/barrel (159 liters or 42 US gallons) which translates to about 42 GJ/tonne: liquid ammonia's combustion heat is 383 kJ/gram mole (17 grams) which translates to 22.5 GJ/tonne and making it via electrolytically generated H_2 requires about 14.2 MWh of electrical energy.

75% of 30.3 billion barrels of oil consumed by 35% heat-efficient engines adds up to 4.85E+19 (0.75*30.3 E+9*6.1E9) Joules worth of energy services (useful work). Providing that much useful energy via "direct electricity" would require the full time output of 1538 [4.85E+19/1E+9/3600/24/365] one GW_e reactors.

If we assume that those energy services were to be provided by engines burning "nuclear ammonia", 6.19 billion [= 0.75*30.3E+9*6.1E+9/22.5E+10] tonnes of it would be needed per year. At 14.2 MWh/tonne, making that much of it with electricity would require the full time output of 9970 [6.19E+9/14.2*3.E+9/1E+9/3600/24/365] one GWe nuclear reactors – over six times as many of them.

The lesson that calculations like these should teach us is that our descendants should electrify as much of their world's transport system (and almost everything else) as possible.

The shipping industry currently emits about 1 gigaton of CO_2 per year, or 2.33% of global emissions, according to Low Carbon Pathways 2050, a "$4 million multi-university and cross industry research project. International trade associations are leading the effort to decarbonize that sector in alignment with the goals set by the Paris Climate Agreement. Their immediate challenge is simple to state but hard to address: "*ambitious CO_2 reduction objectives will only be achievable with alternative marine fuels which do not yet exist.*"

To meet the targets defined by the Paris Agreement, the study determines that the shipping industry would need to achieve "*net zero emissions by approximately 2035 (1.5°C) and 2070 (2°C).*"

Its key findings conclude that the industry needs to do two things: first, act swiftly and, second, identify a viable carbon-free liquid fuel [84].

- Shipping will need to start decarbonisation soon because as stringency increases over time, increasingly costly mitigation will be required. The later we leave decarbonisation, the more rapid and potentially disruptive it will be for everyone.

- A substitute for fossil fuel will still be required as energy efficiency improvements alone will not be sufficient in the medium to long term.
- Energy storage in batteries and politically correct renewable energy sources will have some role to play, but will likely to still leave a requirement for a liquid fuel source.

While low carbon fuels (bio or synthetic fuels such as ammonia) may be necessary in the timescales modeled in its report to enable international shipping's low carbon transition, under current technology costs they were not deemed economically viable [85].

However, that study assumed that ammonia would only be used in expensive fuel cells, overlooking its potential in combustion-based engines.

The first such use that I'm aware of occurred at the same Norwegian hydropower plant that produced the heavy water (D_2O) which was to be the moderator[28] of Germany's first nuclear reactor during WW II[29]. Before that war most of its electricity had been used to make fertilizer ammonia with electrolytically-generated hydrogen. A bit of that ammonia fueled the truck depicted in Figure 9. Germany's scientists never did get their reactor built but the Norwegians continued to power their truck and fertilize their fields with ammonia.

If France's leadership 30–40 years ago had decided to build enough hdditional reactors to produce its own synfuels as well as electricity, its current "yellow vest" populist uprising would not be happening.

---

[28] A moderator's purpose is to slow down the expremely fast (high energy- 1-2 Mev) neutrons created during fission and thereby increase their probability of capture by other fissile atoms thereby keeping the reaction going. A perfect moderator atom does not "capture" (irreversibly absorb) neutrons, it's just something that they can repeatedly collide with and thereby loose kinetic energy. All else being equal, the lighter the atom, the better it serves that purpose. While hydrogen is actually a better moderator than deuterium, it absorbs too many neutrons to permit a reactor containing natural uranium (0.7% fissile ^{235}U) to reach criticality because too many neutrons are absorbed by non fissile ^{238}U. That's why a light water-moderated reactor's (LWR's) uranium must be "enriched".

[29] Since electrolytic hydrogen gas is somewhat richer in the light isotope (H) than is the liquid water from which it is made, repeated distillations can produce the pure heavy water (D_2O) required to moderate natural uranium-fueled reactors. That's why Hitler's Germany demanded that it become Norsk Hydro's raison d'etre.

Figure 9 Norsk Hydro's ammonia-fueled truck circa 1933.

There's a consensus about the technical challenges that an ammonia-fueled engine must overcome. The Caterpillar Corporation put it succinctly in their 2008 patent application covering a "*power system having an ammonia-fueled engine*": "*When ammonia is combusted, the combustion produces a flame with a relatively low propagation speed. Its low combustion rate causes it to be inconsistent under low engine load and/or high engine speed operating conditions. Most existing combustion engines that use ammonia as fuel require a combustion promoter (i.e., a second fuel such as gasoline, hydrogen, diesel, etc.) for ignition, operation at low engine loads and/or high engine speed.*"

In other words, one option is to use a liquid combustion promoter. However, as Caterpillar's patent subsequently noted, this approach "*generally requires dual fuel storage systems, dual delivery systems, and dual injection systems, thus adding additional weight, complexity, and cost to the engine system.*"

The second option is to use hydrogen as the combustion promoter. A promising variant of this approach is to place an on-board "reformer" between the fuel tank and the engine. The reformer "cracks" enough of the ammonia into hydrogen and nitrogen to support viable combustion. Crackers can be simple in mechanical terms, consisting of a heated chamber furnished internally with a catalyst. However, the patent describes one more challenge: "*The requirement for the combustion promoter fuel fluctuates with varying engine loads and engine speed, which can cause control issues.*" This means that cracking a fixed proportion of the ammonia (or a proportion that varies simply with the rate of fuel flow) is unlikely to produce good engine function.

A third option would be to use it as Honda currently does with the optimally operated small[30] gasoline engines charging the batteries of its hybrid

[30] They can be small because car engines have to provide only 7 to 15 KW (10-20 horsepower) most of the time.

cars. This makes especially good sense in areas that get too cold for "pure" BEVs to function properly (e.g., most of Canada).

There are many ammonia engines in development, including by Caterpillar, a major supplier of maritime, tractor, truck, and industrial engines. In addition, some major industrial projects will soon demonstrate the environmental benefits of ammonia in dual-fuel combustion. If the shipping industry is committed to acting quickly and identifying carbon-free liquid fuels ("*which do not yet exist*," they say), it must evaluate ammonia as a short-term bridge fuel for engines, as well as a long-term energy source for fuel cells.

Ammonia is emerging as an economically viable fuel for dual fuel engines where it is substituted for natural gas [86]. When ammonia is so used, a significant reduction in CO_2, PM (particulate matter) and rather surprisingly even NOx, emission levels is observed. Ammonia is a high-octane fuel (possesses a high resistance to pre-ignition), which means that higher compression, more efficient, engines can be used. Consequently properly modified ammonia-fueled engines exhibit enhanced power output compared to their gasoline or diesel fueled counterparts.

Technavio noted that shipping represents roughly 65% of the market for dual fuel engines, which are "*gaining popularity in the marine industry as a growing number of vessels are using these engines over conventional diesel or gas engines ... enabling the crew to adhere to various marine pollution (MARPOL) regulations regarding propulsion engine emissions.*"

Farm tractors represent another big potential market for such engines because they typically run at "full load", and, because of ammonia's popularity as a fertilizer,[31] much of the necessary fuel distribution infrastructure already exists.

3.2.3 "The Age of Substitutability"

Table 4 lists major raw materials consumption other than water and sand/gravel by the USA circa 2009. Note the dominance of fossil fuels. Goeller and Weinberg's "The Age of Substitutability" approached the problems posed by the future's otherwise inevitable "cheap" raw resource depletion with empirical technical data rather than the then and still dominate economic models. Their paper went through the entire periodic table

[31] Gaseous ammonia injected directly into moist soils is immediately hydrolyzed to ammonium ion which is strongly retained by the soil's cation exchange capacity.

Table 4 The USA's material needs circa 2009.

Material	Mtonnes/a (% Imported)
aluminum	3.25 (45%)
ammonia	22 (45%)
plastics	28 (?)
steel	93 (25%)
cement	100 (20%)
natural gas	403 (19%)
coal	858 (2%)
oil	984 (71%)
Σfuels/total materials	90.12%

examining all the elements plus some of their more important compounds[32], to determine humanity's demand for them circa 1970, estimating for each the total resource available assuming a broad definition of potential sources – the atmosphere, the ocean and the uppermost one mile-thick layer of the earth's crust – not just "rich" ores. The ratio of total resource to demand was determined for each element in terms of years until exhaustion – a measure of their relative abundance or scarcity. Throughout that list, in cases where there was a clear indication of a finite life time, they identified that element's most important uses and possible substitutes. For example, some combination of abundant titanium, aluminum, and iron could probably serve the same purposes currently served by much rarer metallic elements. Aluminum could be obtained from abundant clays rather than rare bauxite and titanium substituted for exotic stainless steels. Based upon their analysis of geological and technological data, they then pronounced the principle of "infinite substitutability"; i.e., with the exception of phosphorus, mercury, and most importantly, the chemically reduced forms of carbon and hydrogen (coal, oil and gas) serving both then and now as fuel, a rich modern civilization could continue indefinitely utilizing only the Earth's nearly inexhaustible resources. The key to such a future would be the availability of energy cheap enough to wrest

[32] For instance, the fully oxidized carbon within carbonate rock can't be used for the same things as can the carbon in fossil fuels.

valuable materials from low grade ores rather than from much less abundant rich ones.

Another reasonable substitution that Goellor and Weinberg didn't mention would be to switch from Portland to geopolymeric-type cements [69]. Concrete is currently the world's fourth- most human-consumed substance after water, sand/aggregate, and fossil fuels. An average of approximately three tons of it is produced for every person on earth each year. Making the Portland-type cement "binder" utilized in the majority of that concrete accounts for ~5% of current anthropogenic CO_2 emissions, one-half of which is due to the fossil fuels (mostly coal) currently heating most cement kilns.

Since about 3 GJ of heat energy is required to make one ton of Portland cement [87], the ~3.7E+9 tonnes of it currently produced/consumed each year would require ~1.12E+19 J of electricity – equivalent to the full time output of about 355 full-sized nuclear reactors (~1.2% of the total (~30,000) needed to satisfy 100% of our descendants' energy needs).

The substitution of geopolymeric binders (cements) – mixtures of low calcium fly ash and/or calcined clay "activated" by solutions containing ~15 wt% Na_2O in the form of sodium silicate and 45 wt% NaOH – for Portland-type cements would greatly reduce carbon emissions. Even if the sodium hydroxide utilized to make the activators were to be produced by reacting lime (CaO) with sodium carbonate (trona), only about 15% as much limestone would have to be calcined to produce an equivalent amount of finished concrete.

Another plus for geopolymeric concretes is that they are more durable than are those made with Portland or "Roman" (lime-pozzolana) cementitious binders. The reason for this is that the sand/aggregate binding-mineral assemblage formed during their curing process is neither hydrated nor readily carbonated by atmospheric CO_2 and/or bicarbonate-containing water.

Finally, regardless of which binder is employed, the future's concrete infrastructure would last much longer if basalt fiber reinforcing bar/wire were to replace today's steel rebar [70]. Modern lightweight steel rebar-reinforced concrete structures are less durable over the long haul than were those made by the Romans because the carbonation of any calcium silicate based concrete eventually lowers the pH of its pore fluids to a point that allows embedded steel rebar to rust, expand, and thereby crack the concrete surrounding it. Basalt fiber rebar does not rust and is intrinsically cheaper than steel rebar because the majority of the earth's crust consists of basalt and melting/spinning it requires less energy than does iron smelting. It is also ~7 times stronger mass-wise than steel, which means that less of it would be required.

The energy generated by 75 one GWe reactors could produce 1.5 tonne/year of environmentally correct, basalt fiber reinforced, geopolymeric concrete for eleven billion people every year.

Of course those substitutions would be impossible unless some sort of abundant, reliable, cheap, and clean power/energy replace today's finite carbon-based industrial fuels. Of the possibilities consistent with Mother Nature's rules and facts, Goeller and Weinberg concluded that breeder-type nuclear reactors offered the most promise. Unlike the majority of the world's political leaders both then and now, they recognized that over the long haul, decisions based upon technical information (facts) will serve humanity better than would those based upon convenient political and econometric assumptions (alternative facts).

3.2.4 Additional Apps

If the world were to be fully electrified via a properly implemented nuclear renaissance, its air and water would be cleaner, its homes and cities more livable, and far more interesting, better paying, and more secure employment opportunities would be available to its young people because they would be the ones employed building/maintaining their brave new world. Finally, a rarely mentioned (too politically incorrect) reason for China's monumental success during the last four decades is that its leadership adopted/enforced a one child per family policy at the same time they decided to encourage/enable its people to become both creative and entrepreneurial [88]. The purpose of that policy was to free up time and capital which could then be (and was) devoted to "making China great again". It also rendered children born during that period especially "special" to both their parents and society-at-large which in turn rendered their lives more enjoyable and successful. Population growth rate would drop precipitously [9] as would the degree of misery/desperation/frustration currently driving young people everywhere (mostly males) to join terrorist gangs and hate groups. In other words, the world's currently desperately poor people would experience the same benefits of nuclear powered prosperity that Japan's, South Korea's and China's have enjoyed.

A government's job is not just to give its currently most important citizens (its "base") whatever they want, but to pave the way for a prosperous, stable, and safe future for everyone. Any kind of government mandated fertility control is unattractive, but unless its goals are achieved otherwise (e.g., by improving the lives and futures of already-living people), it's apt to become necessary everywhere if wars don't render it moot.

4
Why Sustainability Requires Breeder Reactors

Light water cooled/moderated reactors (LWRs) generate the vast majority of the world's nuclear power – it's become a locked in technology [89].

There are two basic types of LWRs (see Figures 10 and 11): Pressurized Water Reactors (PWRs) in which the normal (i.e., H_2O aka "light") water surrounding/cooling their fuel rods within the core isn't allowed to boil, and Boiling Water Reactors (BWRs) in which it is.

They are otherwise similar in that: 1) their fuel rods consist of zirconium alloy tubes containing enriched (typically 3 to 5% ^{235}U) uranium oxide pellets; 2), they operate at very high pressures (>1000 psi) but low temperatures (~285°C); and 3), they "waste" many of the neutrons generated when the fissile atoms fueling them (mostly ^{235}U) "split". The latter characteristic refers to the fact that in moderated reactors, too many neutrons are "captured" by materials (water, zirconium, fission products, control rods, and "burnable poisons") other than the fertile ^{238}U comprising the bulk of the uranium, meaning that they can't "breed" as much fissile (in this case, ^{239}Pu) as they consume. Consequently, regardless of how small and modular they might become[33], any "advanced" version of today's light water cooled/moderated reactors (e.g., NUSCALE's 60MWe SMR) would consume at least ~160 tonnes of natural U/GW_e/year, most

[33] "Small" and "modular" are about the only "advanced" reactor characteristics that DOE NE's leaders have consistently emphasized/supported for the last 15 years or so. That's why we've ended up with inefficient mini-LWRs (NUSCALEs) instead of something able to "save the world". "Modular" and "small" are fine attributes if seeking them doesn't distract research scientists/engineers from concentrating upon what they should be trying to do. A downside of "small" is that there's about as much overhead cost associated with running a small power reactor as there is a big one. Another is that neutron leakage is higher/GW, meaning that they don't "breed" replacement fissile as well. If the world is to be powered with fission-type reactors, those reactors must be able to at least isobreed. "Modular" is good any way you look at it and many powerful MSR concepts could be modular because their cores would be small.

Darryl Siemer (ed.) Nuclear Power: Policies, Practices, and the Future, (51–58)

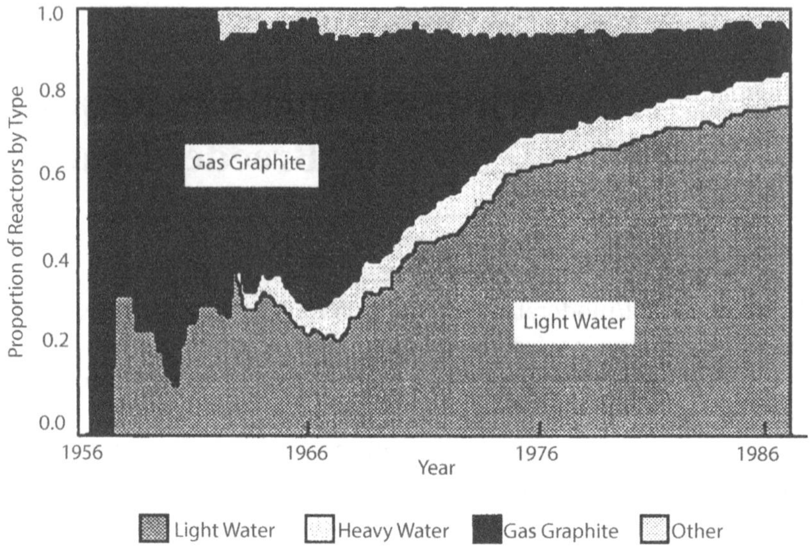

Figure 10 Power reactor types (IAEA 1987).

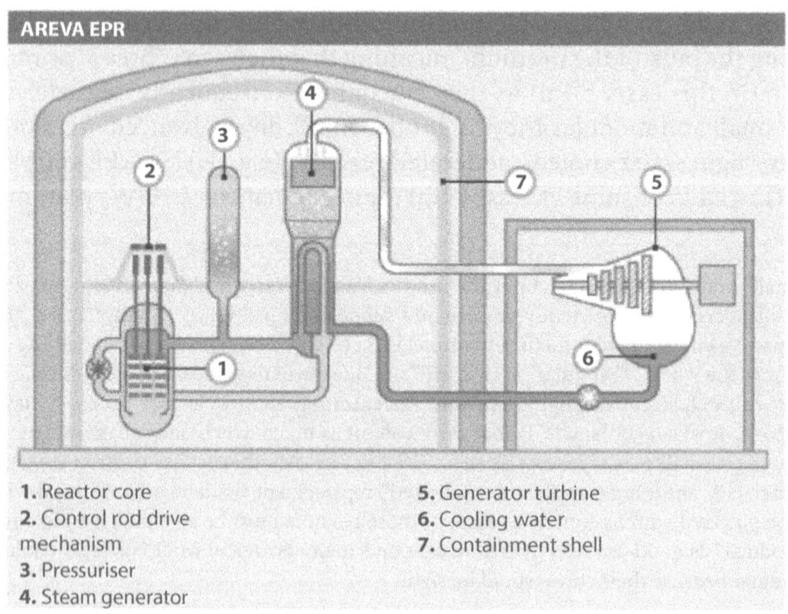

Figure 11 State of the art Gen III^{+} PWR schematic (GOOGLE "EPR").

of which, primarily ^{238}U, would be discarded during its fuel fabrication's U-enrichment step[34]. A genuinely sustainable nuclear renaissance could not be implemented with them or any other sort of burner/converter-type reactor because the uranium industry's official estimate of all affordable (in that context) "proven plus undiscovered uranium resources" is only ~18 million tonnes [90]. Generating 2 kW's worth of LWR-generated electrical energy for 11.2 billion people, would consume 100% of the world's "affordable" uranium within about 5 years [1.8E+7/22,000/160 = 5.1].

People championing such a burner/converter reactor-implemented nuclear renaissance do so based upon economic models rather than geological/technical data; i.e., they assure us that sufficiently cheap uranium will inevitably be discovered when the definition of "affordability" rises beyond today's upper limit, <$260/kg. For example, if a ten-fold price increase would indeed unearth 300 times as much uranium [91][35], a conventional light water reactor-based, 22 TWe, nuclear renaissance could be fueled for several centuries before the cost of its resource-fuel (i.e., natural uranium – NU) exceeded that of coal-fired power plants if the price of coal were to stay the same as it is now. Unfortunately, I see no proof that Mother Nature actually follows any such simple relationship, which renders such assurances unconvincing (see [92]). For example, the uranium industry's latest (2018) Redbook's fuel resource figures do not support Deffeyes & McGregor's rule-of-thumb; i.e., that a ten fold increase in what we're willing to pay for uranium would generate 300 times as much of of it. Plots of that Redbook's $\log_{10}$(resource size) vs $\log_{10}$(cost) data possess slopes of from 1.10 to 1.25 depending upon whether the resource in question is "identified", "reasonably assured", or "inferred". Their rule of thumb implies a slope of 2.477 [$\log_{10} 300/\log_{10} 10$].

The notion that we simply haven't yet bothered to look hard enough for uranium (or thorium) is also questionable because unlike most of the mineral resources that mankind seeks, natural actinides are always accompanied with sufficient mildly radioactive decay products to be easily detected with readily available, cheap, portable, meters.

---

[34] Although most of the natural uranium (NU) discovered/mined/processed to make LWR fuel never ends up in that fuel, all of it must be converted to uranium hexafluoride – the gaseous form required by the world's isotopic separation (enrichment) facilities.

[35] Kenneth. Deffeyes was a geologist who worked with M. King Hubbert at the Shell Oil Company's research laboratory and then became a Princeton University professor. He was the author of the book, *Beyond Oil: The View from Hubbert's Peak.*

The Earth's total surface area is about 510 million square km, ~29% of which is land. Assuming a rock density of 2.8 g/cc the mass of land-sited crustal rock to a depth of 1 km is about 4.14E+17 tonnes. Since such rock contains an average of 2.8 ppm uranium, the total mass of "accessible" uranium (we're already digging deeper than that for oil, gas, gold, and diamonds) comes to about 1.16E+12 metric tonnes or 64,000 times that currently deemed "affordable" if utilized by today's inefficient burner/converter reactor-based nuclear fuel cycle.

The real problem is that the vast majority of the world's uranium is in low grade (~100 ppm) deposits (shales, phosphate rock, etc.) too expensive to first mine and then process to feed a grossly inefficient nuclear fuel cycle (see Table 5).

Another commonly made/heard assertion, that the "reprocessing" of today's ~240,000 metric ton (total actinide) accumulation of spent LWR fuel could address their fuel limitation issues is also wrong[36]. It takes about six spent LWR fuel assemblies to make one "MOX" fuel assembly and that process can only be done once because such fuel's "reactor grade plutonium" is degraded to "crap grade" plutonium that cannot be further burned in the same sort of reactor. This means that the maximum possible such fuel supply extension would be about 17%. MOX fuel is also expensive to make and relatively dangerous/bothersome to use, which is why most independent reviewers and all US electrical utilities have concluded that it's not worth the trouble.

For instance, a 2007 analysis of the pros and cons of reprocessing concluded that a US built reprocessing plant would cost about $50 billion and its operating cost would be $1000–3000 per kg of spent fuel processed [94].

Ignoring build costs and assuming $2000/kg spent fuel processing cost, making 1 kg of fuel for a reactor grade plutonium based MOX fuel would cost about $12,000 (6*2000). Assuming 4.5% enrichment and 0.0015 tails discard during it, making one kg of NU-based fuel would require 8 kg of NU [.045/(.0071-.0015)]. The current cost of NU is about $62/kg which makes the U going into in a kg of such fuel worth about $372. Doubling that figure to account for its enrichment cost, one kilogram of that fuel's "heavy metal" (actinides) comes to $744 or just 6.2% that of MOx-based fuel.

The reprocessing of spent light water reactor fuel makes economic sense only if the actinides so recovered serve to fuel breeder-type reactors[37].

---

[36] Appendix I describes the hows and whys of reprocessing.

[37] However, even at $12,000/kg such fuel costs only about one half as much per kWh as does typical US coal.

Table 5 Average uranium concentrations (parts per million) in ores, rocks and waters [93].

Material	Concentration (ppm U)
High-grade orebody (>2% U)	>20,000
Low-grade orebody (0.1% U)	1,000
Average granite	4
Average volcanic rock	20–200
Average sedimentary rock	2
Average black shale	50–250
Average earth's crust	2.8
Seawater	0.003
Groundwater	>0.001–8

Because any genuinely sustainable nuclear fuel cycle would require the resumption of reprocessing in one form or another, I've written APPENDICES I & II to describe its history and what its future might look like.

Fueling burner/converter-type reactors with uranium extracted from seawater couldn't "save the world" either. The country most involved with testing/developing that scenario is Japan (see [95] for a slide set and lecture). Dr. Tamada's slides begin with the contention usually prefacing such reports; i.e., that "there's 1000x as much U in the oceans as on land". That's incorrect because there's about 350 times as much U in "readily accessible" rock (the first kilometer of the ~3 ppm U crustal rock covering the earth's land surfaces) as in its seawater (~1.33 billion km^3 of ~3 ppb U water). Dr. Tamada's slides then go on to describe Japan's pilot plant scale demonstrations and end with a conclusion that a 68.7 by 15.2 km (1030 km^2) array of the most promising uranium adsorbent developed (amidioxime-coated, irradiated polyethylene fiber "ropes") might be able to collect enough uranium to fuel six of Japan's almost state-of-the-art LWRs (i.e., ~1200 tonnes U/a) for a "reasonable" cost. That conclusion assumed that the adsorbent would be trapping ~4 g U/kg adsorbent per cycle, roughly three times more than was generally recovered during their demonstrations [96]. A subsequent US study concluded that an "improved" version of that adsorbent would capture ~3.3 g U/kg adsorbent [97]. In any case, because those adsorbent arrays must be bottom-anchored, experience significant wave action, and situated where natural currents quickly replenishes the water surrounding them, almost 4 million km^2 of shallow (mostly coastal) ocean bottom would have to be covered to fuel 22 TW's worth of conventional burner/converter-type reactors. Finally, anyone considering such schemes should be aware that both fishermen and the world's "Rainbow Warriors" are apt to object – the former because their nets and lines would surely become entangled by so situated gigantic synthetic "kelp beds", the latter because those arrays might also entangle/strangle seals, whales, and turtles.

Other than for those little technical details, mining the oceans to fuel conventional burner/converter-type reactors is a really fine idea.

On the other hand, coupling Dr. Tamada's uranium filters to a breeder reactor powered desalination plant's brine outlet could collect about 20 times as much uranium than that reactor would consume[38]. This means that if the energy devoted to the future's water desalination systems represents >5% of mankind's total needs, seawater could indeed fuel everything.

Realization of Weinberg and Goeller's utopian (but possible) future can happen only if the future's decision makers decide to first develop and then implement an appropriately scaled (big enough), sustainable, nuclear renaissance and then see to it that untrammeled human nature does not turn that/their project into yet another interminable cost-plus nuclear boondoggle.

A genuinely sustainable nuclear fuel cycle must be implemented with reactors that generate ("breed") as least as much new fuel ("fissile", any readily fissioned actinide isotope) as they consume from "fertile" natural uranium and/or thorium and thereby constitute a "renewable" power source. A non-renewable resource (aka finite resource) is a resource of economic value that cannot be readily replaced by natural means quickly enough to keep up with consumption. Fossil fuel generated power/energy is non-renewable because Mother Nature's fuel-making reactions are many orders of magnitude slower than is the rate at which we are consuming them.

For example, each year the Earth's plant life "fixes" about 7.8E+13 kg of atmospheric carbon of which ~99.9% is oxidized back to CO_2 and

[38] Assuming that the desalination plant processes twice as much seawater as it produces fresh water and that its breeder-type power plant's heat to electricity conversion efficiency is 50%, it would take about 3kWh's of fission heat to process each cubic meter of seawater. 3kWh = 1.05E+7 Joules requiring 3.28e+17 U atoms which ='s 1.3E-4 grams ^{238}U. One m^3 of 3 ppb seawater contains 3E-3 grams of U – about 23 times more than that reactor would "burn".

returned to the atmosphere when those plants die [286]. The majority of the 0.1% that isn't so oxidized mixes with the clay fraction of soils to form organic ("black") shales where it is very gradually converted it to some combination of humic acids, kerogens, peat, graphite, coal, natural gas, and petroleum. 99.5% of the roughly 2% of it that eventually ends up as petroleum remains tightly bound within those shales. Consequently, Nature's overall petroleum-type fuel carbon production rate is about 1.56E+9 (7.8E+13*0.001*0.02) kg/year of which 7.8E+6 kg is as easily/cheaply recoverable as was Texas' petroleum when I was born. Mankind currently consumes (mostly by burning to CO_2) about 93 million barrels of petroleum per day which translates to about 3.7E+12 kg of carbon (='s 13.6E+12 kg CO_2) per year which mostly ends up in the atmosphere: in other words, we're burning petroleum ~475,000 times faster than Mother Nature is creating it.

"It is hard to know which is the more remarkable - that it took 600 million years for the Earth to make its oil, or that it took 300 years to use it up."

M. King Hubbert

Breeder reactors, not today's power reactors, would be renewable because it would be easy/cheap to fuel them for at least as long as Homo Sapiens is apt to exist – hopefully (and very optimistically) another million years.

5

Today's More Promising Breeder Reactor Concepts

5.1 Heavy Water Thorium Breeders

Probably the quickest way to begin the implementation of a suitably fuel efficient (sustainable) nuclear fuel cycle would be to convert today's ~293 already-paid-for pressurized water reactors (PWRs) to breeders[39] by replacing their uranium-based fuel assemblies with uranium/thorium based assemblies and substituting "heavy" water (D_2O) for the "light" water (H_2O) currently used to both moderate and cool them [98]. The reasons to expect much better (more fuel efficient) performance are that deuterium doesn't consume (waste) nearly as neutrons as hydrogen, and ^{232}Th can efficiently breed its fissile (^{233}U) with slow-moving (moderated) neutrons while ^{238}U can't. The same changes would probably also enable the World's ~77 full scale "boiling water reactors" (BWRs) and 32 CANDU-type reactors to achieve breakeven fissile regeneration [99, 100].

Unfortunately, so modified/operated reactors would possess the same weakness of any solid fueled reactor; i.e., in order to run for a long time their fresh fuel assemblies would have to contain extra fissile, meaning that its initial excess reactivity (which would otherwise cause run away heat generation - an "explosion"), would have to be suppressed with control rods and/or burnable "poisons" both of which deliberately absorb (waste) neutrons that could otherwise serve to breed new fissile. This translates to very little fuel burn up per cycle (~10 GWd/t), short refueling intervals,

[39] Fissile breeding (aka fuel conversion) plays a significant role in the fuel cycle of all commercial power reactors. During fuel burn up the fertile material (usually ^{238}U) is converted to fissile ^{239}Pu which partially replaces the "burned" fissile ^{235}U, thus permitting the reactor to operate longer before the total amount of fissile decreases to the point that criticality is no longer achievable. The ratio of new fissile formed to that consumed is called "conversion factor" which is acronymed either CR or BR: If $CR<1$ the reactor is a converter (aka burner); if $CR= 1$, it's an isobreeder; if $CR>1$, it's a breeder.

Darryl Siemer (ed.) Nuclear Power: Policies, Practices, and the Future, (59–86)

and lots of expensive/fussy solid fuel reprocessing and fuel assembly refabrication.[40] "Burnup" is a measure of the amount of heat generated per tonne of the core's total actinides ("heavy metals" or HM) per "burn" cycle. For example, let's assume that a 33% heat-to-electricity efficient, 1 GWe PWR generates 25 tonnes of spent fuel per year of which 20 tonnes was initially uranium (mostly ^{238}U enriched to about 5% ^{235}U)

total heat generated per year =1/0.33*1E+9 J/s*3600 s/hr *24 hr/d *365 d/year =9.56E+16 joules
1 GWd =1E+9*3600*24 = 8.64E+13 joules
"burnup" in the usual units = 9.56E+16/8.84E+13/20 = 55.3 GWd/tonne HM
HM consumed/year = 9.56E+16/3.2E-11 J/fission *235 g/mole /6.023E23 atoms/mole/1E+6 grams/tonne = 1.17 tonnes
Fraction HM burned =1.17/20 = 5.8% (100% burnup corresponds to ~949 GWd/t)

5.2 Liquid Metal Fast Breeder Reactors (LMFBRs)

Another solid fueled reactor concept has already been implemented and more or less thoroughly tested in the US, France, Japan, and Russia. Russia's first LMFBR, the "BR-1", was designed in 1949 and commissioned at Obninsk in 1955 –four years after the USA's tiny "EBR I" was fired up. After that, two more Russian reactors, the BR-5 and BOR-60, were commissioned & tested, both there and at the Research Institute of Atomic Reactors (RIAR) in Dimitrovgrad. These paved the way via an intermediate –sized (1000MWt, 125 MW_e (most of its heat energy was used to desalinate Black Sea water) "BN 350", to Beloyarsk's first almost full-sized

[40] A year's worth of fuel for a typical pressurized water reactor consists of about forty, 4 meter long, fuel assemblies each consisting of ~289 thin-walled Zircalloy tubes containing sintered/pelletized low enriched (typically 3-5% ^{235}U) uranium oxide. Their fabrication currently costs about twice as much as does the uranium going into them. Altogether, a typical power reactor's fuel-related costs sum to ~ $0.008/kWh. Reprocessing such fuel involves chopping the tubes into short pieces; roll-crushing them to release the brittle uranium oxide ceramic pellets within them; dissolution of that ceramic in strong nitric acid; courter current liquid-liquid extraction of that solution with a solution of tributylphosphate (TBP) in kerosene to separate/recover the actinides: back extraction of those actinides into dilute nitric acid, adding a reducing agent (typ. ferrous ion or hydroxylamine) plus more strong nitric acid; and finally, reextracting the uranium back into more TBP/kerosene leaving the now trivalent plutonium (Pu^{+3}) in the aqueous phase.

LMFBR, the BN-600, which has purportedly been generating 600 MW of electricity since 1980. Based upon that success, construction of the bigger BN-800 began in 1984. However, after the USSR's genuinely disastrous and really stupid 1986 Chernobyl screwup,[41] and that socialist republic's subsequent collapse,[42] that reactor's construction was suspended until 2006, after which it was completed and brought up to a minimum-controlled power level for several years' worth of testing. In 2014, it was finally connected to the grid and reached full power (800MWe) at the end of 2016, thereby reclaiming Russia's ownership of the world's most powerful "sustainable" reactor. Whether any of Russia's LMFBRs were ever operated as breeders of anything but weapons-grade plutonium remains unknown because what goes in/out of its fuel reprocessing facilities remains as much a closely guarded secret as it was within the USA's. However, there's no reason to expect that they couldn't operate in that fashion.

On the other hand, the USA's perpetual dithering about what reactors should do has left its last, best, hope of regaining the lead with its own probably even-better designed LMFBR concept in nuclear purgatory for over two decades [101, 102]. GE Hitachi Nuclear Energy's "Power Reactor Inherently Safe Module" (PRISM) is a "small", modular, sodium-cooled fast reactor concept based upon the 20 MW_e Experimental Breeder Reactor (EBR II) operated at DOE's Idaho site (NRTS/INEL/INEEL/INL)[43] from 1964–1994. Each individual 311 MW_e core would be immersed in a pool-type containment vessel and fueled with metallic (not oxide) fuel. They would be inherently safe due to negative power reactivity feedback, large in-vessel coolant inventory, passive heat removal, below-grade siting, and very low (near atmospheric) operating pressure. The U.S. Nuclear Regulatory Commission stated that, "*On the basis of the review performed, the staff, with the ACRS [Advisory Committee on Reactor Safeguards] in agreement, concludes that no*

---

[41] which fiasco involved a totally different, much bigger, and really primitive Russian-designed "RBMK" reactor much like Hanford's very first "production" reactor. Unfortunately, it was both manned and managed by schedule-driven, good team players that deliberately went about setting up about the only situation that could have possibly caused that accident. It wasn't really their fault because the reactor's fatal weakness (graphite tipped control rods) constituted a state secret too important to reveal to lowly "operators".

[42] Which, in light of the world's rapidly deteriorating political/economic situation, is pretty scary and may, eventually prove fatal to millions (billions?) of people if a technologically advanced nation's leadership decides to start another world, not just regional, war.

[43] NRTS =National Reactor Testing Station
INEL=Idaho National Engineering laboratory
INEEL = Idaho National Environmental and Engineering Laboratory
INL= Idaho National Laboratory

obvious impediments to licensing design have been identified." Unfortunately, none have actually been built and most of the people that had worked on developing both them and their electrometallurgical (a.k.a., "pyroprocessing") fuel recycling system retired over a decade ago [102].

5.3 Molten Salt Reactors

Over the long haul the most promising way to sustainably power the world with nuclear reactors would be to switch from solid-fueled to fluid-fueled, molten salt-type breeder reactors (MSRs). The reasons for this include:

- Since their fuel consists of already ionized (ionic) salts, it cannot be damaged by ionizing radiation. The lifetime of solid fuels is limited by radiation damage to both the fuel "meat" and its cladding.
- Their much lower operational pressures (water cooled/moderated reactors operate at >1000 psi, MSRs would operate at <100 psi) translates to a much lessened explosion potential (greater safety) and much lower construction costs.
- Greatly simplified fuel fabrication, clean up, and recycling because "spent" fuel would not have to be dissolved to purify it and no complex/expensive fuel assembly would then have to made from it.
- Higher temperature operation (500–700 vs 250–300 °C) translates to higher heat-to-electricity conversion efficiencies, more compact/cheaper gas turbines, less waste heat, and much more efficient operation in water stressed regions where air cooling might be necessary.
- Easier/simpler/cheaper fuel cleanup/recycle would facilitate the "burning" of the long lived transuranic isotopes (TRU) currently rendering radwaste management especially "controversial."
- Their highly negative temperature reactivity coefficients translates to intrinsically safer operation and superior "load following" capability (as more heat energy is withdrawn from the fuel salt by heat exchanger(s), they produce more heat and vice versa[44]).

[44] Pulling heat energy out of (cooling) a molten salt causes a volumetric contraction of 200-400 ppm/degree C depending upon its composition. Since the internal volume of the core

- Because fissile can be added whenever it's needed, MSRs would not have to be started up with a huge excess reactivity (extra fissile) – again, this would tend to enhance safety[45].

Conceptually, the only things required of a molten salt reactor is that its fuel be brought into a critical configuration within its core and the resulting heat energy removed from it. Thusly they could be built in many different ways, which explains the diverse array of concepts proposed during the past six decades.

Heat removal is critically important for any powerful nuclear reactor. Many ways have been proposed to achieve it with MSRs including:

1. Pumping a liquid coolant such as lead, mercury or a volatile salt directly into the fuel salt so that it can mix with and thereby extract heat from it (extensively investigated and deemed impractical)
2. Pumping a second molten salt or other coolant through pipes immersed in the fuel salt while also pumping it to enhance convective cooling on their outside surfaces
3. Pumping e fuel salt through external heat exchangers

To date only loop cooled and integral-type pumped fuel salt concepts have been considered sufficiently practical to justify further consideration [103]. "Fast" MSR (FS-MSR) concepts are more flexible in that they do not require moderating media within their core and can tolerate much higher fission product buildups translating to lower fuel recycling/reprocessing costs. The fuel salt addition and salt cleanup portions of FS-MSR-based concepts introduce further configurational variability. Fueling options range from directly adding TRU fluoride or chloride salts obtained from a centralized processing facility to the local processing of used LWR fuel into actinide fluoride or chloride salts using close-coupled salt cleanup infrastructure. As of spring 2018, there are at least 13 different companies/organizations ("nuclear start ups") championing molten salt reactor

tank containing it decreases considerably less per degree, cooling increases the amount of fissile within the core which, in turn, increases its heat output. Conversely, removing less heat tends to reduce the reactor's power output.

45 The reason for this is that gradually adding fissile (e.g., pellets of $^{235}UF_3$) to a molten fuel salt slipstream would be simple. That would also make it relatively easy/safe to initially start up an experimental MSR.

concepts. While their potential economic, safety and operating features are extremely attractive, none of them have actually been operated, built, or even licensed yet.

5.3.1 MSFR

One of the Generation IV International Forum's (GIFs) Advanced Reactor [30] potentially sustainable "Gen IV" concepts, EURATOM's EVOL program's "thorium burning" Molten Salt Fast Reactor (MSFR) is especially promising [104]. EVOL started its decision making with the same graphite moderated single fluid (no separate blanket salt stream) MSBR that ORNL had left off with three decades earlier but soon decided that an unmoderated two fluid (separate fuel and blanket) system would likely be superior. Due to the following constraints…

- Not excessively high salt melting points
- High salt boiling temperature (i.e., generate little pressure at the reactor's operating temperature)
- Good thermal and hydraulic properties (fuel = coolant)
- Stable under irradiation
- Adequate solubility of both fissile and fertile components
- Low tough-to-manage/politically charged radioisotope generation
- Reasonable fuel/fertile salt stream cleanup/reprocessing options
- Low solvent salt neutron absorption and scattering cross sections

…both its fuel and blanket salt streams would be a relatively low melting (565°C) eutectic consisting of 77.5 mole% ^{7}LiF and 22.5 mole% total actinide fluoride(s). The actinides within its fuel salt stream would be about 88 mole % $^{232}ThF_4$ (natural thorium – its "fertile" material) and 12 mole % $^{233}UF_4$ (its fuel/fissile). Its blanket salt's actinides would comprise almost entirely natural $^{232}ThF_4$.

Thorium's "breeding" reactions are as follows:

$$^{233}\text{U} + \text{neutron} \rightarrow \text{two fission product atoms} + 2.49 \text{ neutrons } [105]$$

$$\text{One of the new neutrons} + {}^{232}\text{Th} \rightarrow {}^{233}\text{Pa (quick)}$$

$$27 \text{ day half-life } {}^{233}\text{Pa slowly decays to form } {}^{233}\text{U}$$

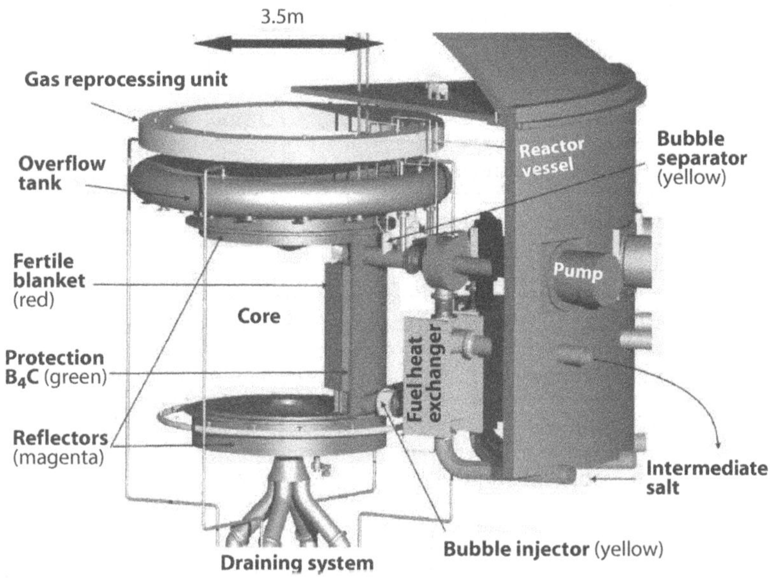

Figure 12 The EU's Molten Salt Fast Reactor (MSFR).

This reactor concept's core is a 2.17 meter "square" (i.e., diameter = height) right circular cylinder surrounded axially by 1-m thick steel reflectors, and radially by a 50-cm thick layer of fertile blanket salt, boron carbide layer, and a thick steel reflector. It is filled with the fuel salt with no core internal structures. The fuel circulates out of the core through 16 external loops, each one including a pump and a heat exchanger. A geometrically safe[46] overflow tank accommodates salt expansion/contraction due to temperature changes. A salt draining system connected to the bottom of the core allows core dumping to passively-cooled criticality-safe tanks to facilitate maintenance or respond to emergencies. This system includes freeze valves that would melt as soon as electric power is lost or the salt seriously overheats. The entire primary circuit, including the gas reprocessing unit, would be contained within a secondary reactor vessel. Fig. 12 is a schematic of its primary loop's layout and Table 6 summarizes its core parameters.

After pondering Dr. Fiorina's presentations at GLOBAL 2013, I wrote an open access paper [106] describing the especially attractive features of an "isobreeding" (generates only as much new fissile (^{233}U) from "fertile" ^{232}Th as it "burns", no extra) version of that concept. Isobreeding constitutes an

[46] Geometrically safe means too small in at least one dimension to prevent so much neutron leakage that any amount of fissile within it could achieve criticality.

Table 6 MSFR Core Parameters [104].

Thermal/electric power	3GWt/1.5GWe
Fuel salt volume	$18m^3$
Fraction fuel within core	50%
Number of heat exchange loops	16
Fuel salt flowrate	$4.5m^3/s$
Salt velocity (0.3 m diameter pipes)	~4 m/s
Blanket thickness	0.5 m
Blanket/volume	$7.3\ m^3$
Boron carbide shield thickness	0.2 m

especially relevant goal/assumption because, at steady-state, the world would not need extra fissile, which fact would mitigate proliferation concerns. Isobreeding would also greatly simplify/cheapen that reactor's operation because very little "reprocessing" of its salt streams (6 to 10 liters per day) would be required to keep it running at steady-state fed with nothing other than natural thorium (see APPENDIX II).

At atmospheric pressure fluoride salts boil at around 1400°C allowing the use of supercritical CO_2 in compact Brayton cycle turbines with thermal-to-electric energy conversion efficiencies ≈ 50%. Fluoride-based molten salts have very low vapor pressures even at red heat, very high volumetric heat capacities (some are higher than that of water at a LWR's >75 atmosphere working pressure), good heat transfer properties, low neutron absorption, aren't damaged by radiation, chemically stable, do not react violently with air or water, compatible with graphite, and some are also inert to common structural metals.[47]

5.3.2 MCFR

Another concept which wasn't formally considered by GIF, the Molten Chloride (salt) Fast Reactor (MCFR), should be simpler to build than the MSFR because its ^{238}U -to- ^{239}Pu "breeding" cycle's superior neutronics

[47] Many people confuse fluoride (fluorine in its "happiest", most stable, & therefore essentially inert state) with elemental fluorine (fluorine in its unhappiest, most reactive & most corrosive state) – they are as different as night and day.

should permit isobreeding without a fertile isotope (^{238}U or ^{232}Th) containing "blanket" surrounding its core. That would significantly mitigate proliferation concerns because at steady-state neither the reactor itself nor its attendant fuel salt cleanup/recycling system would contain "bomb grade" fissile (>90% ^{239}Pu) that might tempt perforce suicidal terrorists to attempt its "diversion". Another of its practical advantages is that its core could be situated within a tank containing either a molten bismuth/lead "reflector" or a molten blanket salt containing a fertile isotope which would allow it to breed startup fissile for other reactors. Like most other MSR concepts, fission-generated heat energy within its fuel salt would be transferred to the working fluid (probably another molten salt) of an external heat exchanger (HX) and then to a secondary HX which then transfers it to its turbines' working fluid (e.g., water for steam/Rankine or carbon dioxide for supercritical CO_2/Brayton). Reprocessing (the recovery and recycle of U & TRU) would entail the chloride-salt based technologies developed for the IFR [107]. Since its reprocessing requirement would be small and sodium comprises most of the cationic material in its raffinates, everything except the TRU, U, and chlorine in the salt so -processed would be vitrified to compact, "best demonstrated available technology" (glass), waste forms suitable for eventual disposal. Its chlorine would be recycled to minimize

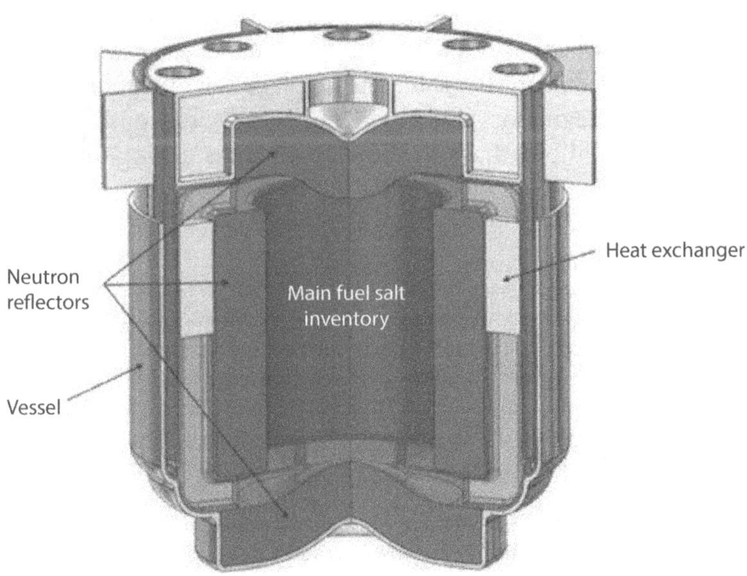

Figure 13 Terrapower's MCFR (https://www.energy.gov/ne/articles southern-company-and-terrapower-prep-testing-molten-salt-reactor).

waste volume, produce better quality waste forms, and prevent escape/dispersal of the otherwise problematic ^{36}Cl.

These reactors would utilize chloride rather than fluoride salts because chlorine (a bigger atom) exhibits much less "moderating" capability[48] and the trivalent actinides serving as both their fissile and fertile isotopes are more soluble in them than in their fluoride counterparts. Natural chlorine's major isotope, ^{35}Cl (76%) gives rise to ^{36}Cl as an activation product – a long-lived energetic beta source, which means that if the system's fuel cycle were to discard chlorine, it would complicate waste management. However, since chlorine discard is both unnecessary and unwise [108], a more compelling reason to use pure ^{37}Cl–based salts is that that ^{35}Cl has greater moderating (via scattering) capability and would therefore reduce fissile breeding. Its transmutation would also generate a little ^{36}S which raises potential corrosion issues[49]. In any case these are probably some of the reasons why the Bill Gates-backed TerraPower "nuclear startup" recently decided to split its reactor development efforts between its solid-fueled, liquid-metal cooled, "breed and burn", "Traveling Wave" concept and an unblanketed, chloride salt-based MCFR, see Fig. 4 [109]. Another startup, "Elysium", is proposing a similar system.

Both of those startups are currently proposing "breed and burn" versions of their MCFRs to render them more attractive to regulators and utility owners. Breed and burn translates to requiring a much simplified (and less effective) fuel salt "cleanup" system – salt seeking FP are not continuously removed[50], which means that the reactor will eventually have to be shut down and that salt either replaced or "reconditioned" (see APPENDIX III). Regulators like breed and burn because it'd be easier to regulate (they would only have to show up en mass during fuel exchange

[48] The "faster" a reactor's neutrons move, the better it can both breed and consume actinide isotopes not fissionable with slow-moving neutrons. Fission products have relatively large neutron capture cross sections in the thermal (slow) energy range but much smaller ones at higher energies. Consequently, much greater fission product buildup is tolerable in an FS-MSR than in a thermal-spectrum (moderated) MSR. Since chlorine is a bigger, heavier, atom than is fluorine, it's a poorer moderator which is why a MCFR would be "faster" than its MSFR counterpart (in either system, halide atoms would outnumber the metals by about 1.8:1).

[49] Such corrosion is unlikely: ^{36}S would likely be either stripped out of the fuel salt as a gas or converted to a non-corrosive and easily removed solid compound via reaction with molybdenum FP.

[50] However, gaseous and noble metal FP "scum" would be more or less continuously removed.

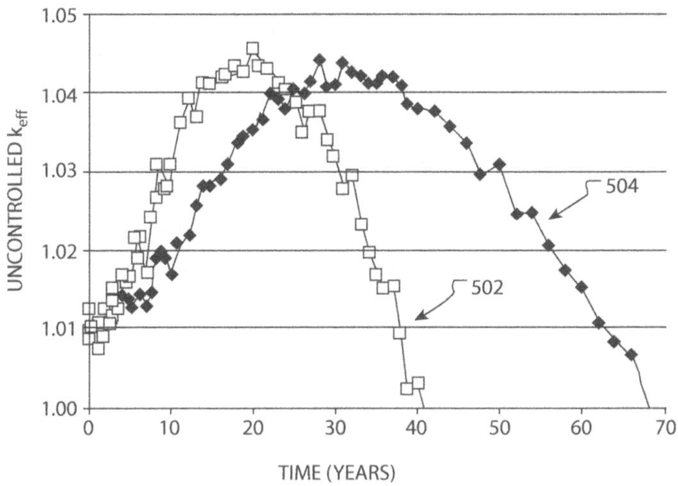

Figure 14 TERRAPOWER's Breed and Burn concept's k_{eff} over time.

periods). Utilities like it because it would simplify reactor operation and therefore probably save them money over the short haul.

My problem with them is that because "breed and burn" compromises reactor performance to simplify its operation, it requires more startup fissile which in turn would limit the rate at which sustainable reactors could be built. Let's go through an example.

This next figure depicts net effective neutron multiplication factor (k_{eff}) ("reactivity") as a function of time of a "breed and burn" MCFR concept described in one of Terrapower's artfully written patent applications[51] [110]. The words characterizing it don't specify its shape (probably right circular cylindrical), size, or how it would be controlled but does disclose the following:

"FIGS. 1A-1F utilizing the 66NaCl-34UCl$_3$ composition. Curve 502 depicts a modeled kÿcurve for a power level of 5800 MW and curve 504 depicts a modeled kÿcurve for a power level of 3420 MW. It is noted that both curves

[51] Terrapower "publishes" only in the patent literature. Like most patent applications and, unfortunately, the ABSTRACTS of many paywalled scientific reports, it's written to claim as much as possible while revealing as little technical information as possible. This section's goal is to demonstrate how such perforce "open access" information can be used draw conclusions about reactor concepts - not determine exactly whatever *Terrapower* (or *Elysium*) is currently proposing. My effort's 65+ cubic meter core volume figure matches that mentioned elsewhere within that patent disclosure (see APPENDIX III for a further discussion).

502, 504 are modeled to operate with a depleted uranium (DU) feed and without specific lanthanide removal. As shown in FIG. 5, the 3420 MW case (curve 504) may operate for nearly 70 years before going subcritical, while the 5800 MW case (curve 502) may operate for approximately 41 years prior to going subcritical. In addition, the model shown in FIG. 5 also predicted a fuel burnup of 43% without any lanthanide removal during the years of operation. Thus, modeling shows that chlorine-based uranium fuel salt may be effective at reducing dependencies of prior molten salt reactors on enriched uranium to maintain criticality."

Another statement in that application says that, "*a fuel salt having a ^{235}U content of 10 mol % may have a reactor core section volume of approximately 67 cubic meters (m^3)*". Another says that the reactor's initial U enrichment level) would be such that its separation wouldn't raise proliferation concerns, i.e., be under 20% ^{235}U.

In order to translate such-written technical information to something meaningful to someone trying to decide whether or not the concept makes sense, it's necessary to combine it with some technically reasonable educated guesses.

Here goes:

First, generating 3420 MWt for 70 years (Fig. 9's curve 504) adds up to 7.55E+18 heat-type Joules [3420E+6*3600*24*365*70].

If 45% of that heat is converted to electricity (reasonable), that's 107.7 GW_e years' worth of electrical energy [7.55E+18/1E+9/3600/24/365].

Since a single fission event generates about 3.2E-11 J (~ 200 Mev) and one gram mole of ^{235}U comprises 6.023E+23 atoms (Avogadro's number) the total amount of uranium consumed during that "burn" comes to about 92 metric tonnes[7.55E+18*235/3.2E-11/6.023E23/1E+6].

Next, I will assume that the reactor 's reactivity is maintained at k_{eff} =1 (i.e., it's "controlled") by adding a 34/66 mole percent depleted $UCl_3/Na^{37}Cl$ mixture which pushes the same volume of its "too-hot" fuel salt into a critically safe overflow tank[52]. I'll also assume that the amount of uranium so added is equal to the amount of fissile consumed via fission, i.e., 92 tonnes.

---

[52] Any reactor's "effective neutron multiplication factor (keff)" must be maintained at almost exactly 1.00 or it'll either "die" (K_{eff}<1) or become supercritical (k_{eff}>1) which could possibly lead to an explosion. Since each individual neutron absorption/fission occurs far more quickly than either human or automated control systems can respond, all practical reactor concepts rely upon natural negative feedback mechanisms (e.g., Doppler broadening & thermal expansion) which work because a fraction of the neutrons ("delayed neutrons") generated per fission aren't immediately released.

If 92 tonnes represents 43% of the total uranium fed to the reactor during its breed and burn cycle, the grand total must have been 214 tonnes (92/0.43) which suggests that that reactor's core must have started out containing ~122 tonnes (214-92) of uranium.

Application of the principle of "additive molar volumes" (see APPENDIX IV and ORNL3913, p 27) permits one to calculate that this fuel salt mix would possess a density of about 3.64 g/cc and that each cc of it would contain about 1.89 g U. This means that the volume of fuel salt within the reactor's core is about 65.6 m^3 [122E+6/1.89/1E+6], which translates to a 5 meter diameter spherical or 4.36 meter "square cylindrical" core.

Next, since it would be impossible for a reactor containing a 34/66 molar mix of trivalent uranium and sodium chloride salts to simultaneously possess "a ^{235}U content of 10 mol % with uranium that's under 20% enriched[53], I'm going to assume that the folks who wrote Terrapower's patent application meant to say that their example's reactor's uranium's enrichment level was 10%.

If initial enrichment is 10%, that translates to 12.26 [0.10*122] tonnes of startup fissile (^{235}U).

If we next assume that the depleted uranium coming out of the enrichment plant has 0.15% ^{235}U in it, the amount of 0.71%^{235} U, NU needed to make12.2 tonnes of ^{235}U fissile would be about 2178 tonnes [12.2/(0.0071-0.0015].

That's 20.2 [2651/107.7] tonnes of NU consumed/GWe which is about eight times [160/20.2] less than that required to generate the same amount of electricity with state-of–the-art LWRs.

If the reactor's initial ^{235}U enrichment level were 20% rather 10%, its first breed and burn cycle would be 4 times more fuel efficient than a LWR's.

It's not a real breeder but is indeed much more fuel efficient than are almost all of today's power reactors (the likely exceptions being Russia's LMFBRs). Additionally, since the reason that it would eventually "die" after 70 years is that it has become choked up with fission product "poisons", its "spent" salt contains more than enough fissile to operate the same reactor once that FP is removed. This means that this reactor does indeed represent something that could be a genuinely sustainable system.

That plus the fact that "breed and burn" would allow decision makers to kick the politically charged "reprocessing" can on down the road for another half century or so makes it pretty attractive.

Again, I feel that their main drawback is that breed and burn reactors would require a fairly large amount of startup fissile. In this example, it

[53] 20% of 34 mole% = 6.8, not 10 mole% ^{235}U.

would be 7.93 tonnes/GWe (12.2/(107.7/70)) which means that going that route would likely limit the rate at which a sustainable nuclear renaissance could replace both fossil fuels and wind/solar/biofuel-based substitutes.

5.3.3 MOLTEX

Another promising MSR concept not considered by GIF is MOLTEX's "stable salt reactor" [111]. Its primary technical distinction is that its core consists of bundles of thin-walled steel tubes containing the fuel salt immersed in a big tank of coolant salt. The fuel salt within those tubes contains a fissile isotope (any combination of trivalent ^{239}Pu, ^{235}U, and/or ^{233}U) in a molten chloride-based solvent salt containing a fertile isotope (typically $^{238}UCl_3$) and table salt (NaCl). Unlike the other concepts I've mentioned, its fuel salt would be "static", not continuously recirculated between its core and external heat exchangers. Fission-generated heat energy would pass through the walls of those tubes[54] to a rapidly upwards moving (pumped) fissile-free surrounding, fluoride-based coolant/blanket salt comprised (in breeder versions) of a ThF_4-NaF eutectic[55]. The coolant stream's heat energy would be transferred to a third molten salt stream (e.g., "solar salt" a low melting (eutectic) mix of sodium and potassium nitrates) which, in turn, would exchange its heat with water or CO_2 to generate the high pressure gas driving its power turbines. Both its fuel and primary coolant salt streams would be rendered non-corrosive to conventional stainless steels via redox buffering with divalent Zr – an extremely powerful reducing agent and fluorine/oxygen scavenger. In this writer's opinion, the MOLTEX concept's chief virtue is that it should be considerably easier and cheaper to implement a "first of a kind" (FOAK) breeding-capable version of it than either the MCFR or MSFR.

The reasons for this include:

1. Building and operating one would not require exotic materials

[54] This concept's "eureka moment" was the realization that in the Earth's gravitational field, a nominally static fuel salt contained within thin-walled (~0.05 cm), ~1 cm diameter steel tubes similar to those typically used in LMFBRs, could transfer their heat to an external coolant sufficiently rapidly [112]. The reason for this is that the density gradients caused by the fuel salt's heat internal generation (~50 kW/liter) would generate a great deal of local turbulence enabling rapid convective (not just diffusional) heat flow from the tube's contents to its walls.

[55] Non breeders would utilize a $NaCl/ZrF_4$ eutectic (minimum melting) coolant salt.

2. It's intrinsically "small and modular" (each 150 MWe/module weighs about 18 tonnes - see Fig. 15) meaning that its parts could be factory fabricated/trucked to a build site and that a reactor could easily be scaled up to produce as much power as needed (stack additional modules side by side in a common coolant salt tank)
3. It's apt to be easier to convince both the government's decision makers and investors that building one wouldn't be too risky

Regarding point # 3, regulators would be comforted by its simplicity (there's nothing mysterious about how it would work and it's not very complicated) and the fact that couldn't possibly suffer a Fukushima-like meltdown (upon shut down, its core would be air cooled via natural convection - no huge water pools and pumps needed). Investors are apt to like the fact that it should be possible to build one both quickly & cheaply.

As far as I am concerned, another significant plus is that its developers are willing to reveal more "technical details" than most of their competitors in today's MSR startup sweepstakes do [112]. Credibility in this hyper-secretive technical arena is tough to earn and roughly proportional to the degree to which a concept's champions embrace "openness".

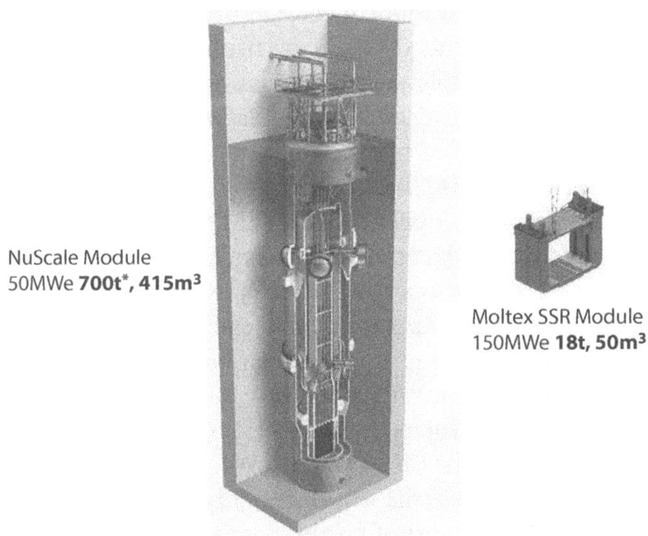

Figure 15 Relative sizes of MOLTEX (MSR) & NUSCALE (LWR) "small modular reactor" modules.

Its primary downsides are its startup fissile requirement (my calculations suggest ~12 tonnes/GWe) and that each reprocessing cycle would require the refabrication of roughly the same number of individual fuel tubes as would an equally powerful LMFBR. However, also like the LMFBR, its high (~150 GWt/day) fuel burnup per cycle means that it would require only about 10% as much reprocessing/refabrication as would breeding-capable (thorium-burning, heavy water moderated) PWR, BWR, or CANDU reactors.

The MSR concepts I've described so far are all "fast" because their cores would not contain a moderating material (e.g., liquid water, beryllium oxide, zirconium hydride, or elemental carbon) to deliberately slow the rapidly moving (fast; i.e., > 1 MEV) neutrons initially generated by nuclear fission. As also previously mentioned (it bears repeating), this enables superior fuel (fissile) regeneration capability, lessens minor actinide (Am, Cm, etc.) build-up, and permits operation with much higher fission product "ash" salt concentrations which, in turn, translates to a much lessened fuel salt reprocessing (clean up) requirement. It also means that their relatively small cores could be "modular" (transportable) and that they would not generate large amounts of radwaste comprised of irradiated/contaminated/damaged moderator. To date, the world's graphite-moderated "production" (of weapons grade plutonium) reactors have generated roughly 250,000 tonnes of radiologically contaminated graphite, most of which lingers in "temporary" storage [113]. In principle, if they were to be operated properly (sustainably), any of them would convert almost all of their actinides, both those introduced and generated *in-situ* via neutron capture/transmutation, to relatively short-lived, simple-to-manage, fission products (FP). All of them would obviate the cost, waste, and safety-related issues inherent to potentially sustainable, solid-fueled reactor concepts such as the sodium-cooled "Integral Fast Reactor" (IFR), "Traveling Wave" concepts or General Atomic's helium-cooled "Energy Multiplier Module (EM2) [114]. All of them should be cheaper to build than "advanced" versions of today's dominant light water reactors because they would operate at much lower/safer pressures and generate more useful (higher temperature) heat energy. And finally, all of them could be started with fuel comprised of the uranium, plutonium and minor actinides (TRU) extracted from spent LWR fuel assemblies which would simultaneously simplify/cheapen the long-term management of such "waste".

Since there isn't enough ^{235}U or spent LWR fuel-derived plus "excess" weapons-grade plutonium in the world (roughly 2400 tonnes total) to start more than 400-600 of any sort of fast breeder reactor, they would have to be configured to breed extra start-up fissile until enough of them had been built to reach steady state (power everything).

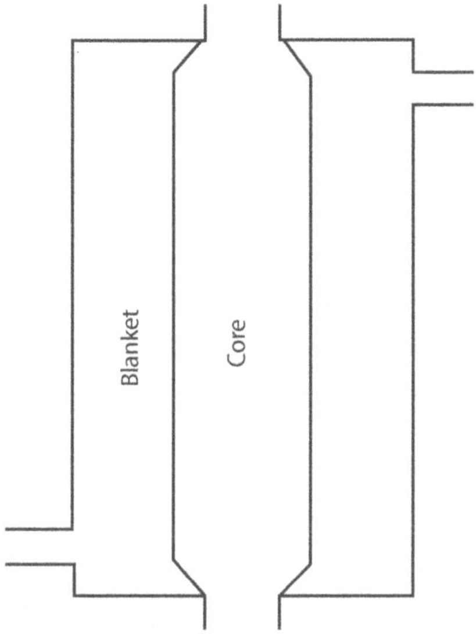

Figure 16 Tube-in-shell two fluid thorium breeder.

5.3.4 Tube in Shell Thorium Breeder

Figure 16 depicts another concept that no one currently seems to be considering that may prove to be the "best" one. It's the "semi fast" (intermediate neutron speed) tube-in-shell configured, two-fluid thorium breeder reactor described by David LeBlanc a dozen years ago [115]. Its internal core (a.k.a., tube) contains a fuel salt comprised of a low melting 2 to 1 mole-wise solvent salt mixture of ^{7}LiF and BeF_2 (FLiBe) containing a surprisingly small amount of fissile (about 0.16 mole% or 6.1E+19 atoms/cc of ^{233}U) and no fertile ^{238}U or ^{232}Th. The outer (a.k.a., shell or tank) side of the system contains a blanket salt comprised of 25 mole % ThF_4 + 75 mole % LiF. Virtually all of its heat-generating fission reactions take place within the core but sufficient of the neutrons so produced pass through the wall separating it from the blanket salt to regenerate at least as much new fissile as is burned within it (the fission of ^{233}U generates two fission product (FP) elements plus an average of 2.48 neutrons[56]. About 1.15 of those neutrons are absorbed by other ^{233}U atoms within the core salt thereby keeping the reaction going and producing a relatively small amount of ^{234}U therein

[56] What revived this concept to me is that, 1) both fuel and blanket salt reprocessing would be simpler than that required by any of the U/Pu based MSR concepts I've seen, and 2) ^{233}U fission apparently releases about 10% more new neutrons than ORNL's researchers assumed six decades ago [105].

while the remaining ~1.33 are absorbed by the system's internals (wall metal, ^{7}Li, Be, F etc.) plus the blanket salt's thorium.

It is stretched-out version of the 4 foot core diameter, 2 fluid reactor (number 36) in Table 3 of ORNL 2751 [116]. It is cylindrical rather than spherical with a diameter (48 inches x 0.78 or 95 cm) relative to ORNL's original concept which emulates its neutronic characteristics (0.78 is the long-cylinder to spherical "geometric buckling factor"[57]). Its core is long enough (7.4 meters) to contain sufficient fuel salt (5.24 m^3) to limit mean heat generation within it to 400 kW per liter, which translates to a whole core heat generation of 2.1 GWt [4E+5*7.4*(0.95/2)^2*1000] or, assuming 48% efficient thermal heat to electricity conversion, 1.0 GWe. Assuming ^{233}U, its withiin-core startup fissile requirement would be 124 kg [=6.1E+19 *1000*5.24*233/6.023E+23] which, if it were connected to a total of 8 m^3 worth of external piping and heat exchanger(s), translates to a total startup fissile requirement of 312 kg [124*(8+5.24)/5.24] – well under 10% of that required by any of the "fast" reactors I've already described[58]. That's a crucially important point because the availability of startup fissile limits the rate at which a big-enough sustainable nuclear renaissance could be implemented[59].

Since this concept's fertile (232Th) isn't mixed in with its fissile (^{233}U)...

1) thorium would not have to be separated from REE FP (U is easy/cheap to separate from REE, Th isn't)
2) since fertile atoms would not be absorbing neutrons within the core, achieving criticality therein would require relatively little fissile

---

[57] A geometric buckling factor relates the shortest possible dimension (usually diameter) of a non spherical reactor to one that is spherical. Spherical vessels possess the lowest possible surface area to volume ratio which means which they minimize neutron leakage/loss.

[58] As is the case with everything else I've written about "other people's" concepts, this section was sent off to Dr. LeBlanc to review and, if necessary, correct. As is also usual, people cc'd on my note seized upon that opportunity to point out that (unlike their own concept), this one is "impossible" because of its inconsistency with today's rules and customs. For instance: "*Main issue is ...proliferation concerning reprocessing of both core and blanket. With the MSBR/LFTR required reprocessing system U fluorination system it is trivial to get pure HEU233 with absolutely NO U232 ... 100% U233 critical mass is about the same as 93% Pu239, and far lower than 93% U235, the currently used weapons main arsenal materials.* "In other words, in that reviewer's opinion the purpose of NE R&D is to devise something that nobody could come up with a hypothesized show stopper based upon the notion that the future's civilian power reactors would be operated by unsupervised homicidal idiots.

[59] This concept's other virtues include compactness and uniquely simple/cheap salt reprocessing.

3) fuel reprocessing/recycling would be relatively simple/cheap (fluorinate out the ^{233}U, distill off/collect most of the FLiBe, vitrify/discard everything else)

Another of this concept's advantages is that its blanket could contain enough fertile Th to effectively shield ^{233}Pa from neutrons until it decays to ^{233}U (the LFTR's relatively small blanket salt volume renders ^{233}Pa separation/storage necessary for breeding)

ORNL published several papers having to do with recovering most of the ^{7}Li & Be from some of its proposed fuel salt compositions via a simple one-plate distillation [117]. It worked fine for that purpose but wasn't of much use for separating thorium from ^{90}Sr and REE-type FP.

One of the virtues of this particular system is that not having Th mixed in with its fissile would render fuel salt clean up/reprocessing much easier/cheaper than it would have been with the MSBR – ORNL's breeding-capable single salt reactor concept [118]: simply fluorinate out/collect the uranium, distill off/collect the FLiBe, & throw away everything else. The remaining waste would be easy/cheap to vitrify and there wouldn't be much of it[60].

Here's another possibly sustainable MSR concept along with two others better suited for niche applications.

5.3.5 LFTR (Liquid Fluoride Thorium Reactor)

The LFTR is a revival of the graphite-moderated, graphite-constructed, two fluid breeder (core fuel salt surrounded by a separate blanket salt) described by ORNL's nuclear experts after it had been concluded that such a system would be virtually impossible to build [119]. The LFTR's chief advocate, Kirk Sorensen[61], has apparently come up with a proprietary (secret) way to get around that issue.

It's not getting much attention for several reasons:

1) It would be a genuinely sustainable, graphite moderated, two-salt MSR which means that its purpose, fabrication and

[60] A GWe year's worth of such waste would contain about 100 kg of precious metals (Ru, Rh, Pd, and Ag) that might be deemed worth separating/keeping.

[61] Kirk has probably done more than anyone else to revive interest in MSRs. Circa 2006 he paid ORNL's document control people to "clear", copy, and then send him its MSR-relevant research reports which he immediately posted on his "Energy from Thorium" blogsite. Those documents along a great deal of other such literature remain there freely available for anyone, "friend or foe", to sift through. Yeah Kirk!

operation would be different than that of any of the reactors adopted/championed by the USA's industry and government afterWeinberg was downsized.

2) Among those decision makers' beliefs is the notion that a reactor's owner/operators should not have to run a close-coupled fuel cleanup/recycling system and manage the waste stream that it would generate. Today's once-through nuclear power systems are extremely simple/convenient in both respects (throwing empty beer cans out of your car's window is also simple/convenient).
3) Since breeding or even breakeven-capable LFTRs would require far more "reprocessing"[62] than would fast thorium-fueled, breeding-capable MSRs, they would probably be considerably more troublesome/expensive to operate than the MSFR or chloride salt-based fast MSR concepts proposed by *Elysium* & *Terrapower.*
4) LFTRs would generate 75-100 tonnes of "crapped up" waste graphite (moderator) per GW_e-year – fast MSR breeders wouldn't[63].
5) As would also be the case with the MSFR, it would be breeding-capable only if its fuel salt were made with isotopically pure ^{7}Li, not natural lithium.
6) As would the blankets of any sort of breeder, its blanket salt would definitely contain "bomb" grade fissile (almost pure ^{233}U) that imaginary terrorists could divert.
7) And, probably most important in today's world, no one like Bill Gates has decided to support LFTR development.

The LFTR's chief virtue is that it if it could be made to work, it would require considerably less start up fissile than would the LMFBR (IFR) or any sort of equally powerful FS-MSR with the exception of Figure 16's tube-in-shell concept.

---

[62] Unlike the "fast" MSFRor tube-in-shell concepts it could achieve "sustainability" only if the 27 day half-life ^{233}Pa generated within its blanket salt were to be continuously separated and stored outside of the reactor's neutron flux until it had time to decay to form new ^{233}U. The reason for this is that ^{233}Pa has a high neutron capture cross section for slow neutrons meaning that if left next to (or within) the core much of it would be transmuted to ^{234}U before it could decay to ^{233}U. Doing so would greatly complicate reprocessing and raise serious proliferation issues.

[63] "crapped up" means heavily radiologically contaminated.

5.3.6 THORCON and IMSR

There are a number of other MSRs currently vying for venture capitalist and government money/attention. From what I've been able to gather, their designs were apparently force fit into compliance with today's light water reactor rules and assumptions resulting in systems little if any more fuel efficient than are today's LWRs[64]. In my opinion while it might be fun to think about how we might go about turning a ballerina into a hippo[65], it is a distraction that the best and brightest among us shouldn't concern ourselves in light of the issues that Hubbert, Weinberg, and Hansen identified a long time ago.

Reactor concepts that couldn't "save the world" would, however, be suitable for some important niche applications; e.g., powering big ships or isolated small/medium sized cities. If the fissile fueling them were to be generated by the breeder-type reactors powering everything else, they could constitute an important part of a sustainable overall system.

One such reactor would be THORCON's "doable" MSR (http://thorconpower.com/). It is a high energy density, graphite-moderated, molten salt converter with a rather short (4 year) moderator lifetime when run at its full power rating. It's "doable" because it emulates ORNL's already demonstrated/proven MSRE [120] in a manner that wouldn't require exotic metals to construct or isotopically pure 7Li to run. A land based THORCON plant would consist of barge transportable 250 MWe modules, manufactured on an assembly line like WWII's liberty ships. Each module would consist of two 400 ton "cans" each of which houses a 250 MWe primary loop including a "pot" (reactor), pump and primary heat exchanger. The cans are duplexed meaning that when one is operating, the other is in cool-down or stand-by mode. The plant is designed so that the change out of a cooled-down can would be both easy and quick. At that time it would be lifted from its silo, moved by the plant's crane to a disassembly area, and those portions of its primary loop requiring replacement remotely separated and stored.

---

[64] The USA's nuclear startups focus upon developing salable IP rather than saving the world. Their "products" of course must be "protected" for as long as possible and I can't really blame them - I used to have to "make a living too" (the good thing about getting old is that you've become more or less comfortably "tenured").

[65] The problem with the "*doing something is better than doing nothing*" argument for designing performance compromised system is that once we humans (esp. businessmen) get used to doing anything in a particular fashion, everyone involved with implementing that/their business model resist change. Since people supplying important things tend to get rich, they also become able to influence politicians to support their business model which, in turn, generates rules, laws and bureaucracies constituting the barriers to science that stifle progress/change.

The baseline THORCON would use a NaF-BeF_2 (nabe) solvent salt mixture for both its radioactive fuel and non-radioactive secondary salt streams because it is already available and reasonably cheap. Since nabe is less attractive neutronically than is a ^{7}Li based salt (FLiBe), it would be only about twice, not 140 times, more fuel efficient than are today's LWRs. The baseline version would employ a tertiary loop using solar salt (a low melting eutectic of sodium and potassium nitrate salts) to remove essentially all of the tritium generated during operation. All the fluid loops within the system are designed to generate natural circulation driven by decay heat in the event of loss of power. Both the secondary and tertiary loops are located in an exterior-to-the-can steam generating cell designed to contain a massive rupture anywhere within these loops. Given a rational regulatory environment, it should be able to generate electricity for between 3 cents and 5 cents per kilowatt-hour.

Another well publicized "doable" MSR reactor concept, Terrestrial Energy's "IMSR" would probably be easier to turn into a ship's engine (https://www.terrestrialenergy.com/). Like THORCON's concept, it is "integral", compact, graphite moderated, and no fertile-containing "breeding" blanket surrounds its core. However, it's more conventional in that the fertile isotope accompanying the fissile (^{235}U) fueling it is ^{238}U, not ^{232}Th, which substitution renders it no more fuel efficient than are today's civilian power reactors.

"Little" reactors like these compare favorably with the huge diesel engines currently powering large ships. The diesel engines propelling the world's international shipping industry currently burn about 298 million tonnes of "oil" per year [121]. At 42kJ/gram, that comes to 1.25E+19 J worth of heat energy per annum. Since state of the art shipping-type diesel engines are a bit more thermally efficient than are THORCONs or ISMRs (~50 vs ~45%), that's equivalent to ~1.39E+19 J's worth of nuclear-type heat energy. When bunker fuel oil costs $600/tonne, that engine's fuel cost works out to about 10.3 cents per kWh or 3–5 times more than that reactor's power should cost. Each 250 MWe THORCON "can" weighs 400 tons or about one sixth as much as a 50% thermally efficient Wartsilla 81 MW engine (currently the world's biggest and most efficient diesel engine). Each such can should be able to continuously power a big ship for about 12 years (250/81*4) which means that a two-can module could probably power it throughout its entire life time. Those savings would really add up: over a 24 year working life time: a 50% thermally efficient 81 MW diesel engine would consume 2.93 million tonnes of bunker fuel, which at $600/tonne would cost $1.76 billion of today's dollars – reducing that cost by two thirds would give that ship's owners/investors a very warm feeling. It would also let them honestly claim that their decision to so power their ship reduced the amount of CO_2 that would have otherwise been dumped into the environment by 9.2 million tonnes. It would also eliminate another environmental impact because bunker fuel is extremely dirty compared to the diesel fuels used in land-based trucks, cars, and tractors, meaning that big ships currently dump more SOx into the environment than do any of Mankind's other fuel burning machinery and/or processes.

Since the 22 TW's worth of breeder-type reactors powering my scenario's clean, green, and prosperous future world would also exhibit about 50% thermal-to-electricity efficiencies, they would be consuming 44 TW's worth of fissile. That's 1.39 E+21 J [44E+14*3600*24*365] of nuclear heat per annum for "everything else" which figure happens to be 100 times that required to power today's shipping fleet with "small" converter-type reactors. This in turn suggests that if the big breeders generated 1% more fissile (CR=+1.01) than they consumed, their extra fissile could power international shipping too.

5.3.7 The Whys, Hows, and History of Sustainable Reactors

All MSR breeder concepts save MOLTEX are at least 50 years old. No potential breeder except the USA's pet solid fueled, "liquid metal (cooled) fast breeder reactor" (LMFBR) technology ever received sufficient attention/funding to generate anything but "paper" (conceptual) reactors with, again, one exception – Dr. Weinberg's/ORNL's graphite moderated "molten salt reactor experiment" (MSRE) which operated for ~four years during the mid-late 1960's [120]. However, that MSR couldn't "breed" for economic reasons[66] - it was too small (8 MWt) and its core was not surrounded with a blanket containing additional fertile ^{232}Th. That R&D program ended circa 1973 because a financially strapped US federal government (the Vietnam war had been fought with borrowed money ballooning its national debt triggering severe inflation) decided to fire the bothersome Dr. Weinberg [122] and "study" only the LMFBR concept because it's a much better bomb-grade plutonium maker. Meanwhile, the USA's nuclear industry was busy selling the first generation of much-enlarged versions of Admiral Rickover's enriched-uranium-fueled, light

[66] ORNL's MSR/MSBR program cancellation in 1973 was accompanied by the "downsizing" of its longtime Director, Alvin Weinberg. The rationale for those actions was that, "*the USA could no longer afford to support two breeder reactor programs*". In 1972 Argonne's LMFBR development work had cost US taxpayers 26 times as much ($123.2M/$4.8M) as had ORNL's MSBR studies - see Linda Cohen's, "The Technology Pork Barrel", Brooking Institution Press, 1991, p. 234.

water cooled/moderated submarine reactor at cost so that it could then profit by servicing (fueling) them thereafter. Although several large LMFBRs were subsequently built and operated in France and USSR/Russia, that concept never gained much traction with utility owners because they already had something that met their needs (LWRs) and didn't want to deal with the problems revealed by the operational history of existing LMFBRs (persistent sodium leaks/fires - [123]) or assume the greater costs of both the reactors themselves and the fuel recycling systems required for sustainable operation.

The main reason why genuinely sustainable nuclear fuel cycles still don't get much attention in the USA is that most of its current political leaders consider nuclear power to be just a stopgap while, with the "temporary" help of fracked natural gas, it "transitions" to an imaginary clean/green world powered by conservation, biofuels, magically modified dams, wind turbines and solar panels/towers (plus some super batteries?) linked together with a zero loss, world-wide magic electrical grid [124, 125]. Consequently, the US government's nuclear scientists/engineers have been helping their "industrial partners" render today's unsustainable nuclear fuel cycle more attractive for temporary niche-filling by championing "small modular" versions of their current reactors even though they understand that such systems don't represent a stand-alone solution to mankind's energy conundrum[67].

Another reason why the development of a genuinely sustainable nuclear renaissance still gets short shrift is that doing so is impossible without first performing the "hot", hands-on research (not paper studies) required to develop anything capable of addressing the future's energy/economic/environmental issues. 750–900 kg of fission products would be generated within any fast MSR's fuel salt per GW_e year, all of which would have to

[67] Here's the ABSTRACT of a recent critique of that policy: "Small modular reactors (SMRs) have been proposed as a possible way to address the social problems confronting nuclear power, including poor economics, the possibility of catastrophic accidents, radioactive waste production, and linkage to nuclear weapon proliferation. Several SMR designs, with diverse technical characteristics, are being developed around the world and are promoted as addressing one or more of these problems. This paper examines the basic features of different kinds of SMRs and why the technical characteristics of SMRs do not allow them to solve simultaneously all four of the problems identified with nuclear power today. It shows that the leading SMR designs under development involve choices and trade-offs between desired features. Focusing on a single challenge, for example cost reduction, might make other challenges more acute. The paper then briefly discusses other cultural and political factors that contribute to the widespread enthusiasm for these reactors, despite technical and historical reasons to doubt that the promises offered by SMR technology advocates will be actually realized" [126].

be properly dealt with to prevent possible corrosion damage[68]. Similarly, the wall/walls of the reactor's core would experience very high neutron bombardment rates that might also cause damage. Answers to the questions raised by these facts cannot be obtained until realistic tests are done under realistic conditions. Because the USA's national laboratories chose to downsize their "hot" experimental facilities/capabilities and replace them with modeling/modelers, it's become extremely difficult/expensive to perform such work – most NE-related "research" funding is currently spent upon shiny new office buildings and personnel-related overhead.

The choice of which sort of MSR to develop/build/use will ultimately depend upon those characteristics deemed most important to the future's decision makers. For instance, if it's still absolutely verboten to have uranium enriched beyond 20% anywhere within a power reactor at any time, they would have to be much bigger (contain more uranium) than would be the case if higher enrichments are allowed. Similarly, if there can't be any sort of genuinely "bomb grade" fissile anywhere within the system, reactor cores will have to bigger and require more fissile to compensate for the fact that they cannot be surrounded by a blanket[69]. If more or less continuous salt-seeking fission product removal (steady state operation) is considered either too troublesome or too potentially "proliferative", it could be put off for long periods by, again, increasing both the size of the reactors and the amount of startup fissile in them (breed and burn). Unfortunately, because startup fissile limits the rate at which a sustainable nuclear reanaissance could be implemented, all of these alternatives amount to kicking the same can further on down the road.

Finally, and most important, successful implementation of any of the sustainable concepts mentioned above would satisfy 100% of Mankind's power needs "forever" with abundant and readily accessible natural

[68] Redox control is necessary because actinide atom fission randomly generates a variety of two-atom fission product atom pairs which may or may not possess the same net stable oxidation state as the actinide. For instance, if the fission of a U+4 atom (atomic number 92) happens to generate a cesium atom (atomic number = 55) and rubidium atom (atomic number 92-55=37) both of which are stable only in their +1 oxidation state, the difference between their before/after net oxidation state (4-2) will cause two of the anions originally accompanying that uranium atom (in this example that's fluoride) to give up two electrons in order to maintain overall electrical neutrality. That would convert harmless, non-reactive, fluoride anions into one of the world's strongest oxidants (fluorine atoms) & therefore most corrosive species. A salt-soluble redox buffering agent (e.g., U+3 or Zr+2) would scavenge up those fluorine atoms producing non-corrosive fluoride salts (UF_4 or ZrF_4).

[69] The fresh fissile created in blankets containing fertile isotopes.

actinide fuel – not just the 0.71% of natural uranium (^{235}U) fissionable in today's moderated reactors[70]. That would render clean/green nuclear power as "renewable" as is sunlight as well as a heck of a lot cheaper, less environmentally impactful, and more reliable [127]. For example, assuming 50% heat to electricity conversion (molten salt reactors run much hotter than do LWRs and would therefore generate roughly 50% more electricity per heat joule (GOOGLE "Carnot efficiency"), just the U within the topmost kilometer of the Earth's continents (i.e., 2.8 ppm of 4.2E+17 tonnes ≈ 1.2E+12 t U), could continuously generate 22 TW_e for 74 million years. Since the earth's crust contains 3 - 4 times as much thorium as it does uranium, let's say a total between them of 12 ppm, breeder reactors could potentially generate ~2.7E+12 J's worth of heat energy from one cubic meter of average crustal rock – about 56 times more than that provided by an equal volume of pure "banked" (*in situ*, not chunked) bituminous coal.

Anyone who feels that such mining would be too environmentally impactful should ponder the following realities:

- The USA's mountaintop-removal approach to coal mining
- The fracking of the relatively "easy" (shallow) 1.6 to 3.2 km-deep shale deposits [128] currently being mined for both oil and gas, causes lots of mini earthquakes, occasionally pollutes ground water, and usually leaks some of its GHG product directly into the atmosphere
- The ~100 ppm U in any shale so mined represents about 4000 times more clean energy as does its ~5% oil/methane/kerogen's dirty energy. This means that4000 times less rock would have to be so disturbed if the right rock fraction were to be "burned" to produce energy
- Huge amounts of phosphate rock containing a similar percentage of uranium have already being mined and processed – the resulting waste piles are already on the surface of the earth and therefore easily accessed.
- Brazil's much heralded recent offshore oil discoveries are even deeper and covered with both seawater and rock [129].
- Nuclear power is far less damaging to both the environment and people than is the coal industry [130]

[70] In reality only a fraction (typically 5/7) of the 235U in NU is usually recovered during enrichment because it's cheaper to do so. However, *in situ* breeding of 239Pu generally generates enough new fissile to compensate for that loss which is why state of the art LWRs require the mining of "only" about 160 tonnes of NU/GWe year.

- The amount of mining required to fuel a breeder reactor-powered world would be much less than required to access the materials required to build a nominally equally(?) capable, clean/green and politically correct (non-nuclear) energy system
- Solar and wind power emits considerably more radiation per kWh than does nuclear power due to the mining/processing (mostly in China) of the rare earths going into them [131]
- According to the U.S. Fish and Wildlife Service, California's Ivanpah "concentrated solar power" (CSP) solar tower-type power plant[71] kills approximately 28,000 birds each year when they instinctively try to perch/land upon its highest points – the tops of the towers at which its ~350,000 huge mirrors are focused

One of the Ivanpah CSP's selling points was that it is purportedly able to provide electricity when the sun isn't shining because when the sun does shine, its molten salts can be heated to 1,000 degrees F and stored in heavily insulated heat energy "batteries" (tanks). However, since sunlight provides only about 23% of that facility's rated 392 MWe capacity, the rest is provided by burning natural gas both there and elsewhere which dumps about 560,000 tonnes of CO_2 per year into the atmosphere. Like windmill-type "capacity", CSPs are big natural gas consumers due to their low efficiencies plus the need to "heat things up" with gas every day before sunrise (if the molten salts were to freeze they could damage its plumbing). And also, since 2013, Ivanpah's owners have twice sought permission to use even more gas than was allowed under the plant's certification agreement – 1.4 billion cubic feet in 2016 [32]. Not surprisingly, solar investment in Spain has dropped by 90 percent from its 2011 level, and worldwide interest in CSP is falling fast.

Windmills also kill lots of birds (especially eagles - GOOGLE it) and bats because the tips of their giant propellers move much faster than birds/bats can fly.

[71] Ivanpaugh was built with the fervent support of the Sierra Club for $2.2 billion including a $1.6 billion federal loan guarantee. Just its build, not its build+maintenance+profit margin, *solar* energy cost averaged over 20 years works out to 13.9cents/kWh – 4-5 times that of today's nuclear reactors.

6

Economics: The Main Reason that the USA's Nuclear Power Industry is Now on the Ropes

Due to Asia's recent rapid economic development, Mankind's total energy consumption grew by 50% during the last 20 years and our appetite for still more continues to grow. However, how that appetite is to be satisfied remains as much of a question now as was at the dawn of the nuclear age seventy years ago. We're just as aware of the finiteness of fossil fuels as we were then, but there are now almost 3 times as many people to feed, and some of humanity's "technical nerds" are beginning to notice that many of the promises made by the folks selling us windmills, biofuels, and solar energy aren't likely to come true. Daniel Yergin's book, "THE QUEST: Energy, Security, and the Remaking of the Modern World" [132], does a fine job of explaining why the tsunami damage to an improperly sited Japanese nuclear power plant caused Germany and several other European powers to declare a moratorium on new plants. Even France, the world's largest exporter of electricity generated by nuclear power, voiced some misgivings about it circa 2014[72].

Germany which has decided to phase out nuclear power entirely invested \$580 billion in renewables and, with the help of its neighbors (primarily Sweden, Norway and France), has achieved 38% green energy. However, its electricity prices are now Europe's highest while its carbon emissions have remained flat due to the fact that it had to burn more coal and imported (mostly from Russia) natural gas to make up for its loss of nuclear capacity. Similarly, the state of Vermont shut down its only nuclear reactor in 2014 and switched to gas which raised its per capita GHG emissions by about

[72] However, those misgivings have evaporated since Germany's recent admission that it will have to buy even more "outside" power as it shuts down more of its nuclear (100% of them by 2022) and coal-fired power plants – consequently, the future for nuclear power in France remains bright.

Darryl Siemer (ed.) Nuclear Power: Policies, Practices, and the Future, (87–118)

5% during a time when overall US emissions declined by 17% due to electrical utilities switching from coal to natural gas fired power plants.

Almost a decade ago President Obama said, "*To meet our growing energy needs and prevent the worst consequences of climate change, we'll need to increase our supply of nuclear power. It's that simple. This one plant, for example, will cut carbon pollution by 16 million tons each year when compared to a similar coal plant. That's like taking 3.5 million cars off the road.... On the other side, there are those who have long advocated for nuclear power—including many Republicans—who must recognize that we're not going to achieve a big boost in nuclear capacity unless we also create a system of incentives that renders all forms of clean energy profitable. That's not just my personal conclusion; it's the conclusion of many in the energy industry itself, including CEOs of the nation's largest utility companies*". [133]

President Obama was referring to an $8.3 billion federal loan guarantee for two (numbers 3&4 –that site already has 2 older reactors) new Westinghouse 1.117 GWe AP-1000 LWRs which were to be built at Plant Vogtle, Burke GA (projected 2500 temporary jobs during the build then 800 permanent). Environmental groups opposed to the licensing of reactors filed a petition in April 2011 (after a good portion of that money had already been spent) asking the Nuclear Regulatory Commission's commission to suspend the licensing process until more was known about the Fukushima nuclear accident. In February 2012, nine environmental groups filed a collective challenge to its design certification and a month later filed a challenge to the license that the Nuclear Regulatory Commission (NRC) had issued. In May 2013, the U.S. Court of Appeals ruled in favor of the NRC. By 24Sep18 that project's projected cost had ballooned to $27 billion resulting in its being dropped completely – the affected utilities simply couldn't afford to keep fighting an uphill battle.

Vogtle's number 3&4's fate demonstrates how costs can quickly escalate when cost drivers are poorly managed and/or the lack of a readied supply chain, excesive litigation, slow regulatory interactions, excessive regulator billing rates, etc., become prohibitively burdensome.

Western-leaning countries' increasing politicization of the relevant rules and regulations combined with large "renewable energy" subsidies, has rendered nuclear power a poor investment and thereby crippled it (Figure 17). Between 2013 and 2018, seven US existing nuclear power plants were permanently shuttered, and 12 others are scheduled for closure through the mid-2020s. The USA is shutting down already-paid-for and well-functioning reactors because their owners can't make money in its now mostly "privatized" electricity markets. Journalists typically translate that fact to the much more attention-grabbing, "*nuclear plants cannot compete*" which,

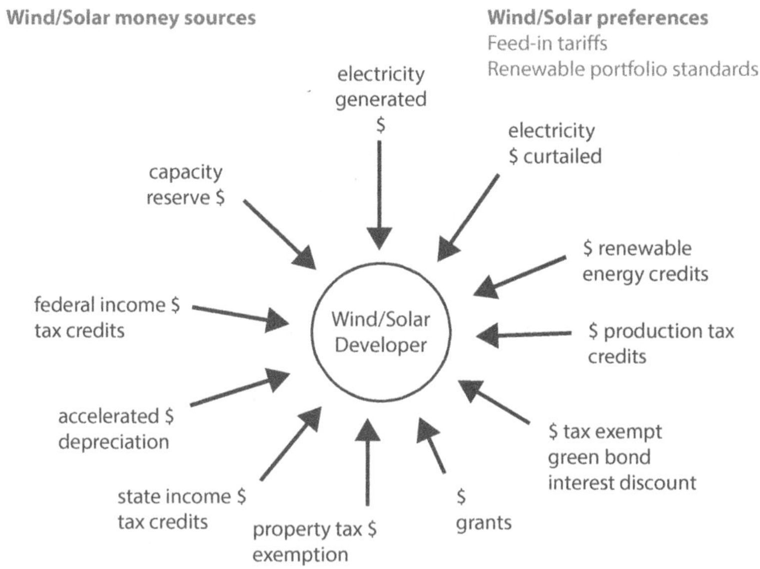

Figure 17 Renewable energy subsidies.

again, happens to be misleading. On a cost per kilowatt-hour of energy delivered to the grid, today's nuclear plants are consistently among the least costly sources available, just above already paid for[73] hydroelectric dams. According to a paper entitled "Nuclear costs in context" [287], the average total generating cost for a US nuclear power plant during 2016 was under $0.034 per kWh, 15% lower than 2012's peak average when Japan's leadership's Fukushima-inspired "decapitated chicken frenzy" had already added additional overhead costs to nuclear power generation here too.

The world's existing nuclear plants were designed and built back when electric utility companies constituted legal monopolies with both an obligation to serve and prices established via a government-approved cost-plus rate of return. In that respect they were much like the "socialized medicine" systems currently serving most of the EU's citizenry. When they were built, no one considered a "what if" in which the USA's electricity supply business would be deregulated and more politically correct and therefore heavily subsidized new competitors might be able to give away their electricity in order to capture additional market share when Mother Nature was cooperating with them – today's situation. This of course means that

[73] In most of the world hydropower is already "maxed out" because most of the good locations for big, powerful, dams are already so used and Mother Nature only provides so much water to fill their reservoirs.

nuclear power "can't compete" in so privatized electricity markets during such times. I also suspect that no one back in the good old days ever considered a "what if" in which nuclear power would not receive a tangible (monetary) "carbon credit" for being green with respect to GHG emissions. In 2016 an analysis performed by the Brattle Group concluded that granting zero-emission credits to nuclear power could secure its economic viability in competition with subsidized renewables and low-cost gas-fired plants. It said: "*A typical revenue deficit for a vulnerable nuclear power plant is around $10/MWh," which is equivalent to costing "the avoided* CO_2 *emissions... between $12 and $20 per ton of* CO_2*, varying with the regional fossil fuel mix that would substitute for the plant*".

Figure 18 depicts California's "dispatchable" power requirement vs time of day during the decade while its citizens were installing a tremendous number of solar panels. Note the following:

- Historically (pre solar) California's total power demand during a typical spring day varied about 25% (from about 18 GWe to 24 GWe) ~ 80% of which was provided by thermal (nuclear and fossil fueled) power plants
- As more and more solar panels were added to the system, the total amount of dispatchable energy (but not "capacity") required was somewhat reduced
- However, more, not fewer, dispatchable power plants had to be built/used because demand maxes out just as the sun is going down in the evening

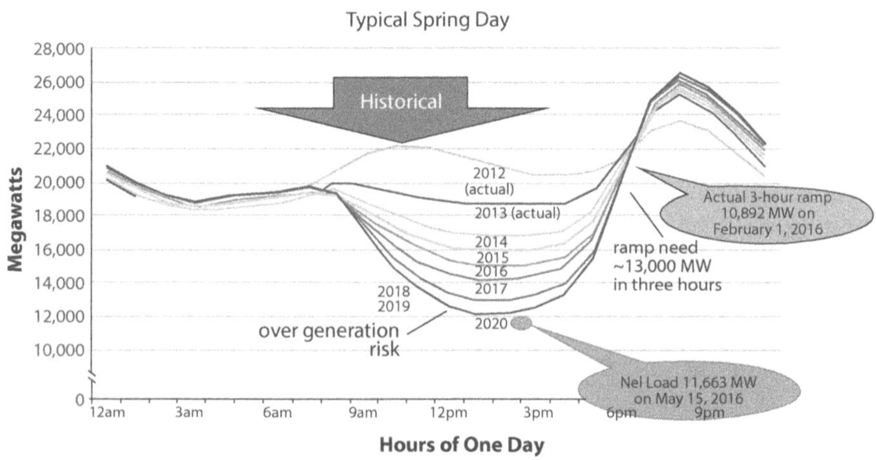

Figure 18 California's 'duck curve' [134].

- Dispatchable power plants still have to supply 100% of demand for about two thirds of each day, especially when it's cloudy
- They must also be able to respond very quickly to take up that load when the sun goes down or the clouds roll in; i.e., respond much more quickly than did California's "old" power system.

Not depicted in this figure is the fact that California's deregulated wholesale electricity spot prices often go negative at the bottom of a sunny day's "duck curve", typically at about 1 PM.

When lots of solar capacity is added to a market, the spot price of electricity will drop precipitously near the middle of a clear day when there is high solar input. The incremental cost of solar production then is about zero and producers receiving subsidies proportional to production will bid electricity prices down to negative values as long as their subsidies exceed them in order to continue sales. The same thing happens on windy days when there's lots of highly subsidized wind turbine power.

In order to address such situations, NE Professors Charles Forsberg (MIT) and Per Peterson (UC Berkeley) devised an especially clever concept[74] (see Figure 19) featuring a big conventional (air Brayton) gas turbine receiving its raw heat energy input from any combination of…

- Heat in the air going into the gas turbine added to outside air by passing it through a massive, well insulated "FIrebrick Resistance-heated Energy Storage" (FIRES) heat energy storage system.
- Heat exchange to either that or "outside" air from heat energy generated by a ~700°C, 236 MWt, fluoride salt–cooled, high-temperature nuclear reactor (FHR)
- Additional "peaker" heat generated by adding/burning natural gas to that air

[74] I highly recommend reading their paper (it's open access) because it is very well written and demonstrates that the USA's best and brightest are still capable of devising a solution to almost any technical problem posed by its business models. However, since it would be almost impossible to render a solid-fueled, graphite moderated, FHR genuinely sustainable, its development should have lower priority than that devoted to the more promising concepts described earlier in this chapter.

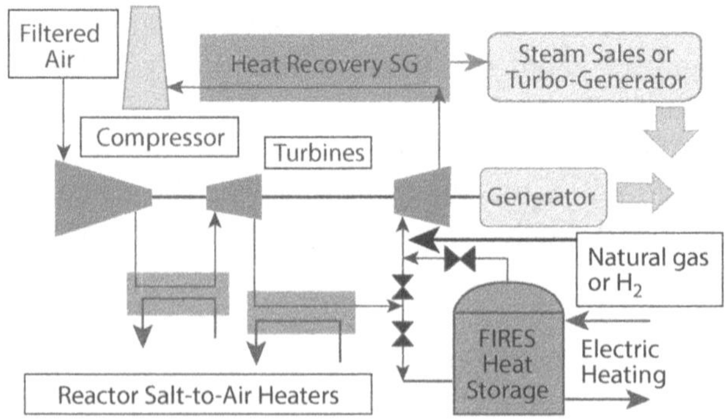

Figure 19 Air Brayton Combined cycle molten salt cooled reactor.

In baseload configuration, no gas is burned and the turbine runs at ~670°C producing ~100 MWe with purely nuclear power (42% thermal to electric efficiency). When demand rises, gas would be added/burned to raise the turbine's air's temperature up to ~1065°C at which point its output would reach its maximum rating, 242 MWe. Finally, when power demand fell below 100 MWe, the difference between it and 100MWe would go to heat its FIRES "battery" via resistance-type heating elements. Gas burned during its peaking mode would exhibit a heat-to-electricity production efficiency of 66% which is about 10% higher than that of a stand-alone combined cycle gas fired power plant[75].

The reactor itself would be superficially much like a pebble bed-type HTGR in that its fuel could consist of golfball-sized graphite spheres containing tiny TRISO kernels slowly circulated through the coolant salt pumped through its core. Unfortunately, although molten salt cooled reactors do possess some of the virtues of true MSRs[76], they also possess the rather special drawbacks inherent to TRISO/HTGR-type fuels; i.e., would be uniquely difficult to implement any sort of sustainable fuel cycle with because reprocessing would be so difficult.

Most of today's solid fueled nuclear reactors could and do vary their output (load follow) at the rate that the western world's total electricity demand changes but operate more reliably and efficiently when run

[75] A "combined cycle" gas fired power plant directs the hot gasses coming out of its gas turbine through a tube-in-shell boiler to generate steam that powers a closed-coupled steam turbine. They are considerably more fuel efficient than "peaker plants" but cost more and can't load follow nearly as well.

[76] For example, their core could/would be much/more compact than a HTGR's because its working fluid/coolant, FLiBe, is a better coolant than is any gas.

steadily at their full-rated power. If they are forced to drastically reduce their output in response to today's (not the past's) deregulated "market forces" and remain in that state too long, a short-lived gaseous neutron absorbing isotope ("poison"), xenon-135, builds up inside their hermetically sealed solid fuel rods which shuts them completely down for a day or so. Molten salt reactors would be better load followers because all gaseous fission products would be continuously sparged (removed) from their fuel and therefore couldn't accumulate.

Being shut down and unable to generate power during market-driven demand transients is more detrimental to a LWR's revenue generation than is simply accepting low or negative pricing whenever the sun happens to be bright and/or the winds especially strong.

In spite of this, during 2017, the USA's ~100 nuclear power plants with a rated generating capacity well under 10% of its total, supplied almost 20% of the electricity actually produced and sold. The average capacity factor of its plants exceeded 92% and the total electrical energy generated/sold was ~800 billion kilowatt-hours. However, their owners continued to lose money, which is certainly not a "sustainable" situation as far as they are concerned – that's why they are responding by shutting down already paid for reactors. Of course, this means that their output has to be replaced with inefficient, natural gas-fired, "peaker plants" that dump lots of CO_2 into the atmosphere whenever Mother Nature isn't cooperating with the alternative energy entrepreneurs[77]. That's perfectly OK with Mr. Trump and his enablers because, to most of them, coal is "clean", fracked natural gas and oil will last forever, and global warming is just another hoax perpetrated by Democrats, socialists, tree huggers, and foreigners.

Contrary to commonly held/expressed opinions, nuclear power is not nearly as heavily subsidized[78] as are either today's favored renewable energy sources or fossil fuels. On March 13, 2013, Terry M. Dinan, senior advisor at the Congressional Budget Office, testified that federal (not the total)

---

[77] "Cheap" peaker plants are fuel inefficient because, 1) they must be kept running (idling) at zero load when wind/solar is able to satisfy power demand and, 2) their exhaust gases don't heat boilers driving additional turbo generators.

[78] With investment subsidies, the financial burden falls upon the taxpayer, while with feed-in tariffs the extra cost is distributed across the utilities' customer base thereby raising everyone's power bills. While investment subsidies may be simpler to administer, feed-in tariffs reward actual production/sales. Investment subsidies are paid out as a function of the nameplate capacity of the as-installed system and are independent of its actual energy yield over time and thereby reward both overstatement of power rating and toleration of poor durability and maintenance. Some electric companies offer rebates to their customers, such as $2.50/watt of "capacity" installed up to $15,000.

energy related expenditure for that fiscal year were follows: Renewable energy: $7.3 billion (45 percent) Energy efficiency: $4.8 billion (29 percent) Fossil fuels: $3.2 billion (20 percent) Nuclear energy: $1.1 billion (7 percent). In addition, he testified that the U.S. Department of Energy would spend an additional $3.4 billion on "financial Support for energy technologies and energy efficiency" as follows: Energy efficiency and renewable energy: $1.7 billion (51 percent) Nuclear energy: $0.7 billion (22 percent) Fossil energy research & development: $0.5 billion (15 percent) Advanced Research Projects Agency—Energy: $0.3 billion (8 percent) Electricity delivery and energy reliability: $0.1 billion (4 percent). That situation has apparently gotten worse: Wikipedia's current entry for "Energy subsidies" ([135] https://en.wikipedia.org/wiki/Energy_subsidies) includes the following figures: Foreign tax credit ($15.3 billion) Credit for production of non-conventional fuels ($14.1 billion) Oil and Gas exploration and development expense ($7.1 billion). The three largest renewable fuel subsidies were: Alcohol Credit for Fuel Excise Tax ($11.6 billion), Renewable Electricity Production Credit ($5.2 billion), and Corn-Based Ethanol ($5.0 billion).

If you go through those numbers adding up everything having to do with rewarding entrepreneurship in renewable, nuclear, and fossil fuel electrical energy generation and then divide by how much energy each actually delivers to its customers (15% for renewables, 20% nuclear, and 65% fossil fuels), you'll see that nuclear power receives about 30% as much of the government's "help"/kWh as do the fossil fuel industries and ~1.8% that given to the people being encouraged to sell us windmills, solar panels/towers, biofuels, etc. This sort of governmental activism has caused electricity prices to spike wherever it is most prevalent (for example, California, see Fig. 20) and also increases GHG emissions because it encourages utilities to substitute natural gas for uranium.

Due to the demand perturbations when additional intermittent renewable electricity is added, standard marginal cost-based pricing approaches tend to generate wholesale electricity prices that are too low for baseload-capable natural gas, coal, and nuclear power providers. Wholesale electricity rates that support baseload capacity tend to diverge further and further from those required to keep them viable as more intermittent power is added to the mix. However, Figure 21's dotted line illustrates that wholesale electricity prices (spot prices) fell throughout Europe, as retail residential electricity prices increased in lockstep with increased intermittent supply. What's particularly interesting is the fact that in 2016, wind &; solar generated 8.3% and 3% of Europe's electricity respectively [136]. This means that what retail consumers have to pay for electricity goes up radically - note

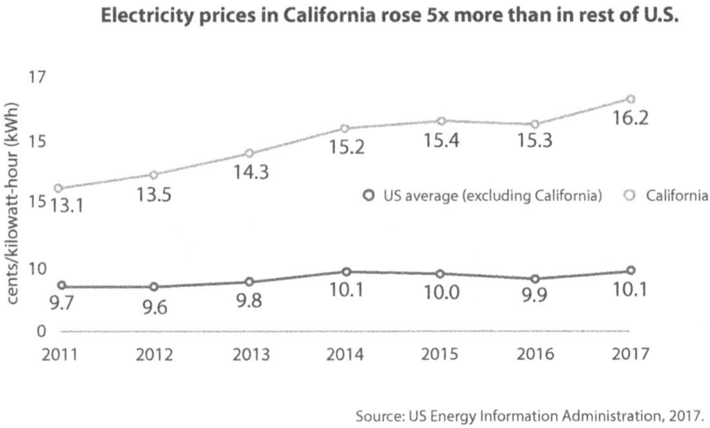

Figure 20 California's "special" electricity costs (volunteered by Alex Canarra).

what happened in Spain (mostly due to its commitment to solar development) and Germany (mostly due to its devotion to wind power) - when even a relatively small fraction of the total power consumed is produced with intermittent sources.

Despite the fact that the consequences of their decisions have been repeatedly pointed out to both US and EU policy/decision makers and

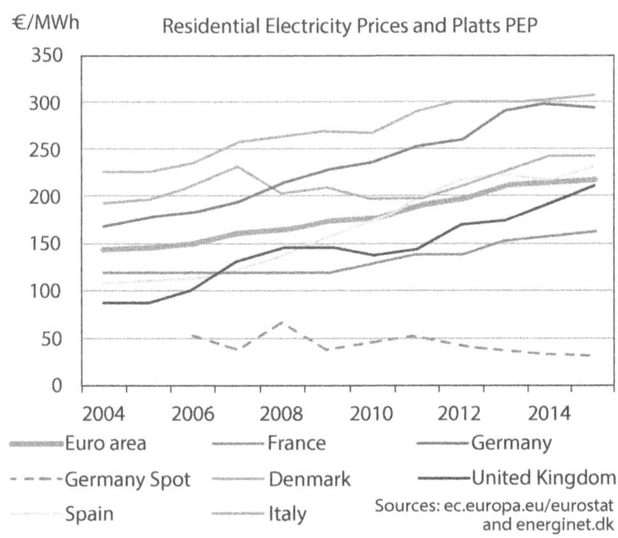

Figure 21 The effects of increasing the proportion of intermittent power on costs.

they usually respond by announcing intent to address them, their/our governments' (state, federal, and sometimes even city) continue to subsidize/support companies installing/operating weather-dependent power sources in areas already subject to frequent periods of over supply when the weather is "right". Though those subsidies are repeatedly scheduled to expire, they are also repeatedly granted last minute extensions which "rescues" are likely to continue due to lobbying - alternative energy companies have become especially politically correct "big businesses" with well heeled owners who employ lots of people and are especially generous contributors to reelection campaigns. In this respect, solar and wind power are much like the USA's deeply entrenched, but energetically and environmentally nonsensical, corn and soybean-based biofuel programs [137].

> "*...on wind energy, we get a tax credit if we build a lot of wind farms. That's the only reason to build them. They don't make sense without the tax credit.*" (Buffet 1014)

The main problem with today's renewable energy sources is not their headline cost per kWh, but that they're a bad match to our demand-driven energy system. ~80% or so of US energy use (after conversions) is non-electric. With the exception of today's already nearly maxed out hydroelectric dams, renewables are non dispatchable electric.

In response to government mandated increases in the proportion of power subject to the whims of Mother Nature (wind and solar), several US utilities are installing megawatt-scale battery systems to mitigate the resulting fluctuations. Utilities in Hawaii, California, and Arizona (APS) have been early adopters and have installed a number of giant lithium ion battery-based "powerwalls" in regions with lots of solar panels.

An April 2018 fire and explosion at a 2 MW battery facility west of Phoenix AZ highlighted the challenges posed by a battery backed up wind/solar based power grid. That particular explosion sent eight firefighters and a police officer to the hospital. The root cause of that and other such incidents both large and small (in airplanes, cars, cell phones, etc.), is that upon recharge the lithium within such batteries tends to form dendrites that penetrate the electrolyte separating their electrodes and short them out. The resulting heat spike ruptures the battery case and its metallic lithium and flammable organic electrolyte quickly ignite & go up in smoke.

Today's electricity storage systems can't render today's intermittent wind and solar energy sources practical. A recent analysis of grid-scale US "unsubsidized and levelized electricity storage" costs [138] for various sorts of storage systems concluded that if lithium ion batteries were to replace today's gas-fired "peaker plants" the overall system's Levelized Costs of Storage (LCOS)[79] would be about \$282/MWh[80]. To put this number into perspective, let's do another calculation: GOOGLEing reveals that the average US household consumes 10.4 MWh per year of electricity costing (retail) about 12 cents per KWh– a total of \$1282/a. If 100% of that power were to be produced with wind turbines and solar panels possessing capacity factors of ~30% and 15% respectively, it seems pretty likely that at least half of the power going into the average US home would be coming from some sort of storage system – not directly from such unreliable sources. If that proportion happens to be 50%, with lithium ion battery-based storage that home's annual power bill would increase \$1644 [50/100*10.4*282] meaning that electricity would now be costing its residents over 26 cents/kWh. The average US wholesale cost of nuclear power is currently about \$0.03/kWh.

As is the case with LWRs, for that application, lithium ion batteries represent a prematurely locked-in technology poorly suited to the problem that it is now supposed to address. However, there are several more promising alternatives. For example, liquid metal batteries can't fail the way that lithium ion batteries do and are also apt to be much cheaper because they feature liquid (molten) - not solid, cheap - not expensive, metallic electrodes separated by a molten salt electrolyte. Their operation relies upon density/SPG differences: a high density (e.g., antimony) metal electrode lies on the bottom of the cell. Immediately above it is an intermediate density molten electrolyte (e.g. a low melting mix of lithium and potassium chloride salts), and a low density liquid metal (e.g. lithium) electrode floats on top. Like oil & vinegar those fluids naturally self-segregate and charging/discharging them doesn't generate the sorts of irreversible physical changes that eventually cause any solid-based battery system to fail (lithium ion batteries are unlikely to last more than ten years in such applications). Professor Donald Sadoway's group at MIT has looked into a plurality of cheap metal (e.g. sodium, calcium, magnesium…) battery chemistries. What makes this sort of battery especially promising is that unlike any of the competing solid electrode based systems, examples have retained >99 percent of their initial capacity after over five thousand, 100% charge/discharge cycles – a

[79] LCOS = sum of all costs per year / MWh stored/discharged per year.

[80] Today's cheapest mass storage option - mass compressed air storage in underground salt caverns - is terribly inefficient. Only about 27% of the energy required to pump air into such caverns is recovered because its heat of compression is lost to the surrounding salt. Consequently that air must be reheated – usually be adding/burning natural gas. https://en.wikipedia.org/wiki/Compressed_air_energy_storage.

characteristic absolutely necessary for a genuinely practical, grid-relevant sized (>>1 MWh) electricity storage system. Professor Sadoway estimates that his batteries will eventually cost about $75/KWh,which, if they last for 20 years, @ one charge/discharge cycle per day, corresponds to a LCOS of just $10.3/MWh [$75/20/(365 cycles/year*0.001 MWh/cycle).

Unfortunately, enough of even such cheap batteries to store one day's worth of 22 TWe power (approximate world demand circa 2100 AD) would cost ~39.6 trillion of today's US dollars.

Edisons'~120 year-old nickel-iron (Ni Fe) battery has proven to be the most durable (by far) solid electrode storage battery. It's extremely robust, tolerant of abuse, (overcharge, overdischarge, and short-circuiting) and often exhibits very long life even if so treated. However, due to its low specific energy/power, poor charge retention, and high cost of manufacture (retail cost per kWh is currently about 60% higher than a TESLA powerwall), other types of rechargeable batteries have largely displaced them. However, it's experiencing a resurgence in some quarters (as the "Battolyser"- GOOGLE it) because overcharging them generates hydrogen which can be stored and used for a host of purposes including energy generation via either combustion or in fuel cells. In that manner, they could accomplish both short-term and long-term energy storage. However, even if they were to last for 30 years, their relatively high up-front cost (currently about $83) translates to a LCOS of $76/MWh.

Upon first glance, some renewable energy technologies currently have energy returns on investment (EROI) higher than that of an equivalent fossil (e.g. new coal) fueled or nuclear powerplant plant. However fossil energy provides the energy required to do the mining, extraction, transporting, building, etc. required to install any sort of politically correct renewable energy source. "*Solar panels are 'rebuildable', not 'renewable'*" [139].

It will take a long time to progress away from today's chaotic "all of the above" approach to supplying electricity and even longer to generate enough of it to meet the needs of a considerably bigger, richer, and fairer future. Over the long haul, the biggest issue facing our descendants will be the unsustainable debt load that we have left them. The USA's external debt - total public and private debt owed to nonresidents repayable in internationally accepted currencies, goods or services – is currently much higher that than any other nation, about 20 trillion dollars or $59,000 per capita. China's external total public and private debt is now about 1.8 $trillion or ~ $1300 per capita. The USA's debt to GDP ratio is now 1.05 - China's is 0.48.

The difference between a country's external financial assets and liabilities is its net international investment position (NIIP). A positive NIIP indicates that the nation is a creditor, while a negative value indicates that it is a debtor. The USA, as recently as 1960 the world's largest creditor, has now become the world's largest debtor. With the recent rapid ascent of Hong Kong Monetary Authority's credit position, China (including Hong Kong and Macau) has recently been jousting with Japan for the top creditor position.

The fact that Hubbert was absolutely right[81] about how long fossil fuels are apt to remain cheap enough to burn combined with the magnitude of the world's debt overhang means that a "green new deal" must be implemented with cost efficient, practical technologies – not with those with especially fine sounding names, eminently politically correct, and/or immediately profitable for well-connected established businesses, entrepreneurs, or investors.

There's currently a good deal of unrealistic propaganda about how tight shale gas and oil could meet our energy needs "for the forseeable future". If "forseeable" means one election cycle or another decade or so, that's reasonable – if it means one average current human lifetime, it's patent nonsense. Claims that society can seamlessly move from fossil to politically correct renewable energy without drastic changes in consumption habits and lifestyles are equally misleading. Most things we're told that sound too good to be true are untrue regardless of their source or possibly constructive intent.

What all this should be telling our decision makers is that it would make sense to develop/build sustainable power plants capable of generating power when it is needed, not just whenever Mother Nature decides to let the wind blow and/or clear away darkness and clouds.

In conclusion, I recently answered this QUORA question, "*How do the economics of wind and nuclear power compare?* with…

> *"This question's answer depends upon what kind of "economics" you're talking about. If you're selling wind mills in a state like California that heavily subsides any sort of politically correct (not nuclear) "clean" power source regardless of how unreliable (intermittent) it is and doesn't care a hoot about its citizens' electricity rates, wind power's economics are really attractive. On the other hand if you're trying to make America great again by building a factory that requires electricity to make stuff, wind power's economics really suck because you can't depend upon being able to actually run it."*

---

[81] Professor Hubbert was absolutely clear about the assumptions behind his conclusions as depicted in this book's figure 3. It's the fossil energy businesses' champions that then cherry picked/changed those assumptions to try to "prove" to the Alfred E. Newmans ("What me worry?") of this world that he was wrong .

6.1 Generic Reactor Build Costs

Another of nuclear power's problems in Western-world countries is their inflated reactor build costs. The fate of Vogtle 3&4 demonstrates how project costs quickly escalate when their drivers are poorly managed and reflect contextual factors (e.g. lack of a prepared supply chain and experienced labor force, slow regulatory interaction paces, expensive regulator billing rate, etc.) representing intractable burdens. In most engineering fields it is possible to accurately estimate a proposed project's cost utilizing information in the "economics" sections of tomes like "*Perry's Chemical Engineering Handbook*". That doesn't work with the Western world's nuclear power projects because even though the facilities themselves aren't especially unique, the costs of everything including labor going into them are both uniquely and artificially high, often by almost an order of magnitude. It is difficult to come up with reasonably accurate build cost estimates because, like those of the USA's health care system, there is little correlation between "should" and "actual" due to sometimes unnecessary, usually self-serving, and often litigious overhead costs[82]. The effect of those cost drivers are best illustrated by a figure (Fig. 22) excerpted from a paper compiling historical construction costs of full-sized civilian light water reactors throughout the entire world [140]. The USA's build cost per GWe started out at under $1 billion 2010-dollars in the late 1960's but quickly ballooned by over an order of magnitude (which eventually killed its nuclear industry) while those of better disciplined countries (Japan, China, France and So Korea) remained at about $2 billion for several decades. Like the cost of real estate, current reactor build costs are determined by "location, location, location" – high in western countries and much lower elsewhere. Consequently, as of 2018, most of the 58 power reactors under construction and 154 planned (220 GWe Total) are in Asia. As of January 2019, another 337 of them have been proposed [141].

Over the long haul, any reactor build cost under ~4 $billion/GWe is economically justifiable because its product will surely be worth much more than that. For example, at 3 cents per kWh, the electricity generated by a one GWe reactor over a 40 year lifetime would be worth $10.5 billion [$0.03/kWh*1E+9 J/s*3600 s/hr*24 hr/d*365 d/year*40 years/3.6E+6 J/kWh].

[82] In the USA it's possible to make a good living doing such things which means that it's become one of the USA's more lucrative white collar "service industries". For example, in 2017 each of the USA's Nuclear Waste Technical Review Board's 14 FTE (full time equivalent) jobs was funded at $257,000 per annum. Each of the The Defense Nuclear Facility Safety Review Board's (DNFSB's) 117 FTE's cost US taxpayers about $190,000.

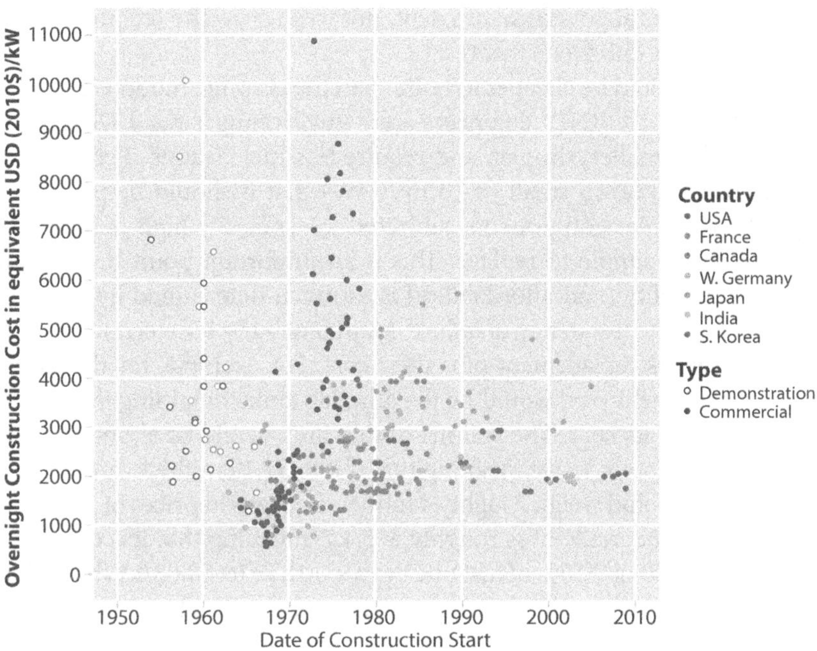

Figure 22 Historic worldwide inflation-adjusted reactor build-costs.

6.2 Sustainable Reactor Build Costs

In 1970, a paper written by two of ORNL's senior-most nuclear engineers [142] included an analysis of what power generated by a full-sized molten salt breeder reactor should cost. Their figure included the costs of construction ($0.159 billion 1970 dollars) fueling, and operating a one GWe breeder reactor. Applying the USA's subsequent ~6.46x inflation factor for most things to their conclusion ($0.0041/kWh) generates a should-cost figure of $0.0265/kWh – about 40% of the USA's current average wholesale electricity cost. This suggests that if the USA's ~320 million citizens set out to power themselves at the rate I've assumed for the future's 11.2 billion people (2 kW_e average with a 40% peak) with one GW_e molten salt breeder reactors, building the 857 [30000*0.32/11.2] of them so required should cost about $860 billion of today's dollars [857 *6.46 *$0.159 B], To put that figure into perspective, it is only about 35% [860/634 =1.35]) more than the USA's current annual "discretionary" (nominally) military spending [143]. Similarly, since the USA's national debt is now ~$22 trillion and the interest rate on its 10 year treasury bonds is ~2.49% (1Apr2019), it seems likely that its taxpayers are paying someone (?) ~$1.5 billion per day just to service their federal government's debt, not to address the technical issues threatening their children's futures.

Fast MSRs should be cheaper to build than the graphite moderated LFTR-type reactors that ORNL's engineers were envisioning circa 1970 because they would be smaller, simpler, and require less fuel cleanup. For instance, their cores could be so small (8-10 m^3/GWe) that it should be possible to design them so that components subjected to especially high neutron flux would be cheap/simple to replace. This is an important point because any product's durability (and affordability) is as much determined by its maintainability as by its frequency of failure. Maintainability is determined by the difficulty of parts replacement plus their cost. For example, for the MSFR, the key replacement part would be the annular tank containing the blanket salt surrounding its core. The volume of that tank would be about 7 m^3 and its surface area about 50 m^2 meaning that if it were 1 cm thick and made of 8 g/cc metal, it would weigh roughly 4 tonnes. Real world prices of the "super alloy" it's apt to be made of is roughly $25/kg, meaning that its cost should be about $100,000 or ~10% of just one day's worth of its $0.03/kWh product.

The secret to making nuclear energy inexpensive has been known ever since France proved that it could be done over 40 years ago. One design is licensed for a generation and then standardized copies built across the grid. South Korea is apparently still able to do so now. In that fashion, all of the logistics plus an experienced construction workforce is ready for every order that comes in[83]. I'm also convinced that THORCON's "build 'em like ships" philosophy is the right way to do things. However, it should be easier/cheaper to build and ship the "guts" of a MSFR, MCFR, or MOLTEX than an equally powerful graphite-moderated MSR like THORCON because they would be considerably smaller and weigh much less. ORNL's single salt MSBR's core contained about 300 tonnes of graphite [118] and the "almost breed and burn" denatured (more politically correct) MSR concept ("DMSR") that succeeded it [144] contained about 2400 tonnes of graphite. Calling any such system "modular" or "transportable" requires the same sort of faith-based reasoning responsible for the outcome of the 2016 US presidential election[84]. The fragile parts of a FS-MSR or tube-in-shell type reactor core would be much more compact and consist of 2–4 tonnes of a

[83] The loss of logistics capability and an experienced workforce is largely responsible for today's nuclear build-cost overruns.

[84] "*Thoughts and prayers*" don't solve technical problems. However, I can sympathize with the "deplorables" who apparently concluded that any sort of change would probably be better for them than more of the business-as-usual thinking exemplified by the previous administration's (& Congress's) "all of the above" energy policy.

"hard metal" that could be quickly crushed into durable half cubic meter waste forms whenever they were removed/replaced.

Unfortunately, utilities are now leery of building any sort of nuclear plant because they represent long term investments suitable only for a stable (predictable) economic system. The stability required to implement a nuclear powered future would require cooperation at all levels including government leadership able to look beyond the next election cycle, a better educated populace that doesn't readily succumb to the whims of fear or fashion, and investors that understand that a rising tide should lift all boats, not just theirs. The lower that interest rates for building nuclear reactors become and the longer those investors can feel confident that their money will keep coming back to them, the lower that energy prices can be for everyone. I don't mean just "competitive with natural gas", I mean low enough to make every other energy source uneconomical.

If we could build a one GWe breeder/isobreeder for $2.5 billion as can South Korea their much bigger and intrinsically more expensive CANDU reactors with 2% (interest) dollars then that facility could either sell its electricity $0.024/kWh for 80 years or first for $0.068/kWh for 10 years and then $0.0165/kWh for the remaining 70 years and still give its investors a 2% annual return. Unfortunately many western societies now can't seem to do such things because the people who must work together to implement any sort of sustainable nuclear renaissance do not evince the necessary maturity, courage, and foresight. Nuclear isn't "too expensive" because of its real costs, but because our leaders don't act like leaders should.

6.2.1 Materials

6.2.1.1 Concrete, Steel, etc.[85]

Generating a kWh's worth of electricity (energy, not "capacity") with any sort of nuclear reactor requires far less building materials than does doing so with today's politically correct power sources (Figure 23).

Let's go through an example assuming that we wish to replace Vermont Yankee's medium sized (620 MW_e) 46 year old LWR with 310 MW (average) wind power, 155 MW (average) PV solar power, and 155 MW (average) of Concentrated Solar Power (solar tower/mirror combo).

[85] The following example's numbers are from Chapter 10 of George Erickson's, "Unintended Consequences: The Lie that killed millions and accelerated climate change", web version is free at http://www.tundracub.com/htmls/unintendedconsequences.html. A more detailed compilation of build costs for GHG-free electrical energy sources may be found in "Critique of 100% WWS Plan" [145].

US Electric Energy Basics

Materials: Fig. 10.4, DoE 2015 Quadrennial Review: **https://tinyurl.com/y8xecbz2**

Table 10.4 Range of materials requirements (fuel excluded) for various electricity generation technologies[52] **Tons/TWHr**

Materials (tons/TWh)	Generator only		CF=.91		Upstream energy collection plus generator CF=.31 & .22			
	Coal	NGCC	Nuclear PWR	Biomass	Hydro	Wind	Solar PV (silicon)	Geothermal HT binary
Aluminum	3	1	0	6	0	35	680	100
Cement	0	0	0	0	0	0	3,700	750
Concrete	870	400	760	760	14,000	8,000	350	1,100
Copper	1	0	3	0	1	23	850	2
Glass	0	0	0	0	0	92	2,700	0
Iron	1	1	5	4	0	120	0	9
Lead	0	0	2	0	0	0	0	0
Plastic	0	0	0	0	0	190	210	0
Silicon	0	0	0	0	0	0	57	0
Steel	310	170	160	310	67	1,800	7,900	3,300
			930			10,260	16,447	

Key: **NGCC** = natural gas combined cycle; **PWR** = pressurized water reactor; **PV** = photovoltaic; **HT** = high temperature

Figure 23 Materials requirements for different energy sources.

Building the solar and wind power facilities would require:

~450,000 tons of steel (0.6% of current U.S. total annual production).
~1.4 million tons of concrete (0.2% of US production/a).

The CO_2 emitted in making the wind/solar renewable energy facilities would be about 2.5 million tons.

The cost of building that much wind/solar generation capacity would be about 12 billion dollars.

The land required to build that much wind/solar capacity would be about: 73 square miles (larger than Washington DC).

Here's a more reasonable suggestion:

Replace Vermont Yankee with a new Westinghouse/Toshiba AP1000 producing 1.1 Ge, almost twice Yankee's.

When we normalize the AP 1000's output to Vermont Yankee's, generating 620 MW's worth of nuclear reactor power would require…

~5800 tons of steel (about 1% as much as would the wind and solar alternatives).
~93,000 tons concrete (about 7% as much as would wind and solar).

CO_2 emitted by the processes going into building that reactor [making its concrete and steel] would be ~115,000 tons (~5% that required to build an equivalent amount of wind and solar energy generating facilities).

Cost: We won't know this for sure until the Chinese finish their first-of-a-kind (FOAK) US and French Gen III+ reactors (AP1000 and EPR), but it is likely that so designed/Chinese fabricated LWRs will cost from 3 to 4 \$billion/GW_e – pretty high but only one third that of this example's wind/solar alternatives.

Another material resource that's much more efficiently used and less impacted by nuclear rather than wind and solar energy power plants, is the land itself. For instance, the area covered by the 4.7 GW_e Fukushima Diachii power plant is about 1.6 km^2 which, assuming 90% CF, works out to about 2700 watts/m^2 of the land so utilized.

Assuming a capacity factor of 0.3, properly spaced wind turbines generate an average of 0.3 watts/m2 of landscape so blighted – about ten thousand times under that utilized by Fukushima Diachii (calculation based upon https://www.theengineer.co.uk/issues/january-2011-online/wind-turbines-need-to-be-farther-apart-suggests-study/.)

Assuming my Nigerian example's solar panels and my home town's (Idaho Fall's, ID) solar insolation value (1891 kWh/m^2), solar panels would generate an average of about 41 watts/m^2.

Another thing that decision makers should consider is that nuclear power plants are far less apt to be destroyed by the world's increasingly frequent/severe floods, tornados, and hurricanes than either solar panels or windmills http://www.theweatherjunkies.com/single-post/2017/09/28/Puerto-Rican-Solar-Farms-Heavily-Damaged-By-Hurricane-Maria.

6.2.1.2 *Other Metals*

The future's molten salt breeder/isobreeder reactors must be made of materials able to retain sufficient physical strength at temperatures up to about 700°C. Those materials must also be sufficiently resistant to high neutron fluxes and the corrosive components of their salt stream(s). A great deal of experimental work performed at ORNL between circa 1955 and 1973-4, suggested that for the fluoride salt-based concepts that it was considering, a molybdenum/nickel/iron/chromium alloy, "Hastelloy N", would likely work[86]. Subsequent Russian and French work agreed with that conclusion. Providing that the redox state of their salt streams were to be buffered at a highly reducing level

[86] The people designing the THORCON reactor plan to use the five fold cheaper 316 stainless steel.

with something like Zr^{++}, MCFRs could probably be built of the same relatively cheap martensitic "HT 9" stainless steel that Argonne/INL concluded would be suitable for its LMFBR's fuel assemblies[87]. Neither metal is particularly difficult to work with nor fabulously expensive (not "unobtanium").

6.2.1.3 *Isotopically Pure Salts*

The most expensive component of any breeding-capable fluoride salt-based MSR like the LFTR or MSFR, is apt to be the isotopically pure ^{7}Li that its fuel and blanket salt streams must be made of[88]. Although ORNL produced a half tonne of pure ^{7}Li during the early 1960's to test its "molten salt reactor experiment" (MSRE), the USA no longer possesses the ability to make it[89]. However, it's very likely that suitably motivated entrepreneurs could produce it today for under \$300/kg [146] which means that all of the ^{7}Li within a MSFR would cost about as much as one day's worth of its product.

Similarly, the most expensive single component of a fast chloride salt, breeding-capable MSR other than its startup fissile would be the isotopically pure ^{37}Cl that its salt stream(s) should be made of. No one makes it now because there's currently no market for it, but, again, it should relatively easy/cheap to produce because the same centrifuges currently enriching uranium, could probably enrich chlorine (in the form of gaseous

[87] Another possibility that I haven't heard about yet is that it seems that the core "tank" of a MSR would be a natural candidate for "cathodic protection". Cathodic protection stops the corrosion of a metal's surface by making it the cathode of an electrochemical cell. One way to do it connects the metal to be protected to a more easily corroded "sacrificial metal" which acts as the anode (sacrificial metal corrodes instead of the protected metal). For other structures an external DC electrical power source provides sufficient current to protect the surface. Investigating ideas like these will require a MSR-type test reactor.

[88] Lithium's other natural isotope, ^{6}Li, has a very high neutron absorption coefficient (cross section) which simultaneously "kills" the reactor's reactivity while generating lots of troublesome tritium. Tritium is both exceptionally labile (can penetrate many metals) and is, in principle at least, biologically impactful. Chlorine-35 exhibits a fairly substantial scattering cross section for fast neutrons which would significantly reduce a MCFR's breeding capability. It also would generate politically incorrect ^{36}Cl (which is also, in principle, biologically impactful) along with some ^{36}S that might be corrosive.

[89] ORNL's "COLEX" lithium isotopic separation process involved electrolysis of aqueous salt solutions with mercury electrodes, which in that first nuclear era's "relaxed" regulatory environment, resulted in serious ground water and soil pollution. Today that separation would likely be done with either atomic vapor laser isotope separation (AVLIS), a method by which sharply tuned lasers are used to separate isotopes via the selective ionization of metallic vapors, or liquid-liquid extraction with crown ethers.

HCl) more easily/quickly/cheaply than they can uranium[90]. It would also be easy to recycle because the same process that represents the "best" way to treat/solidify a MCFR or MOLTEX reactor's waste streams-iron/aluminum phosphate glass vitrification – quantitatively separates chlorine in an easily captured/recycleable form (HCl – see [108]).

6.2.1.4 *Other Materials*

Most thorium breeder reactor concepts would require large amounts of fluorine (e.g., the MSFR's fuel and blanket salts would contain about 33 tonnes of it). Although there's plenty of cheap and readily available fluorine "ores" (fluorite), a good source of already-separated fluorine would be that in the roughly 1.2 million tonnes of "depleted" uranium (DU) hexafluoride stored in/around the world's uranium enrichment facilities. Like the Hanford Site's reprocessing waste, it is a corrosive liquid currently "temporarily" being stored in steel casks, some of which are leaking. It could be converted to safer-to-store and potentially useful solid UO_2 plus useful/recyclable HF via "steam reforming" with some added hydrogen gas.

6.2.1.5 *Startup Fissile*

Since building a GW_e's worth of any sort of nuclear power *except* the aforementioned tube-in-shell thorium breeder would require 4 to 15 tonnes of startup of fissile (^{233}U, ^{235}U, and/or ^{239}Pu), getting enough of it together to start enough breeder reactors to power my scenario's clean/green future constitutes its toughest technical issue.

Before describing a way to address it, I should point out that the current generation (GEN III) of non-breeders also requires that much fissile/GWe to start up – the difference being that they must be continuously fed additional fissile derived from additional natural uranium throughout their entire lifetimes.

Appendix V is a simple program like those we were all empowered to write back when personal computer operating systems included a straightforward, customer-programmable, BASIC language[91]. I've assumed the scenario generally adopted by the world's nuclear engineering (NE) R&D experts for any sort of fast reactor; that is, they would be started up with fissile "pyroprocessed" from spent LWR reactor fuel (e.g., GNEP) plus that in the world's

---

[90] the ease of a centrifugal separation is determined by the square root of mass ratios: It'd be easier to separate chlorine isotopes because [(1+37)/(1+35)]^0.5 >> [(6*19+238)/(6*19+235)]^0.5.

[91] Downloading DOSBOX & figuring out how it works (it's not too tough) lets you use/write your own little programs like this one.

"excess" weapons grade ^{239}Pu and ^{325}U[92]. Since doing that is both technically feasible and serves two politically correct purposes[93], there's a pretty good chance that it might actually be employed. APPENDIX V's example program assumes that Mankind continues to do nothing for another five years (pretty safe assumption) while the necessary R&D is being done while ~400 GWe's worth of LWRs continue to operate. Starting then those LWRs operate for another 15 years while adding new breeder reactors at a rate equal to total fissile "pyroprocessed" from spent LWR reactor fuel[94] plus that in the world's "excess" weapons grade ^{239}Pu and ^{235}U divided by the number of tonnes of startup fissile required per reactor divided by 15. After twenty years additional new breeders are added with fissile (^{235}U) from annually mined natural NU (another variable) plus that generated by the already-built breeders. When the total time INPUTed (NUM which for 2100AD ='s 81) is reached, the program stops looping and prints out how many reactors exist and how much uranium had to be mined/processed to get them all started.

For example, let's assume that we want to build 30 TWe's worth of sustainable nuclear power plants by 2100 AD. If we start building them beginning with 1900 tonnes of spent fuel RGPU + excess weapons grade fissile and each start up requires 5 tonnes of fissile, we'll have about 807 total GW_e of nuclear power (400 LWRs and 407 breeders) at the end of twenty years. From then on, the amount of freshly mined NU required to start 30,000 reactors by 2100 AD depends upon their CR – if they are isobreeding (CR=1.00), nearly 23,000,000 tonnes of fresh NU must be mined/processed. If they make 20% more fissile than they consume (e.g., CR =1.2, the figure attributed to the Russian BN-350), only 7,000,000 tonnes of NU would be required. The former figure is about 25% greater than the uranium industry's 2014 Redbook's total "*proven plus undiscovered resources*".

Such calculations reveal that it is indeed possible to become 100% sustainably nuclear powered with such reactors by 2100 AD but only if we

---

[92] The world currently has about 200,000 tonnes of spent fuel in retrievable storage. Assuming that such fuel contains an average of 1.2 wt% Pu 70% of which is fissile (^{239}Pu+^{241}Pu) that total comes to 1680 tonnes. The USA also purportedly still possesses ~600 tonnes of cold war-generated, weapons grade HEU, [147]. In addition, the USA's production reactors generated roughly 100 tonnes of bomb-grade plutonium (>90% ^{239}Pu), much of which is currently being stored in vaults at DOE's Savannah River National Laboratory. This adds up to a grand total of 2380 tonnes of startup fissile if our leaders decide to beat their swords into ploughshares.

[93] i.e., it would destroy both weapons grade fissile and "reactor grade" plutonium thereby rendering today's stockpile of spent fuel "waste" easier/simpler/cheaper to dispose of.

[94] APPENDIX I goes through some examples of how spent LWR fuel could converted to fuels compatible with breeding capable MSRs.

start soon and quit "burning our seed corn" (^{235}U and ^{239}Pu) in today's or any other sort of inefficient nuclear reactors[95].

Again, the exception to this would be LeBlanc's tube-in-shell reactor concept because it should require far less startup fissile – likely under 2 tonnes ^{235}U - to get a GWe's worth of isobreeding reactors up and running (depends upon reactor size/shape). If that turns out to be the case, then building my scenario's 30,000 full sized breeders would take far less time and NU; i.e., it could probably be completed by 2060AD utilizing the same amount of "new NU" as would continuing to run today's ~400 equally powerful LWR's (see APPENDIX VI).

In this respect, it's a *good* thing that the USA's approach to radioactive waste management is dysfunctional – much of the fissile inventory required to initiate a sustainable nuclear renaissance is now considered "waste" – the management of which received a great deal of "study" resulting in actions/ programs that didn't work out as planned/hoped/promised. For example, SRL's MOX boondoggling turns out to be a *good* thing – its 34 tonnes of "excess" almost pure ^{239}Pu still exists in a useable condition and almost everyone who worked on that project is richer than he/she would have been if no such work had been available (Uber & Lyft can employ only a limited number of the USA's downsized nuclear engineers and scientists).

Simultaneously reestablishing the USA's "leadership in nuclear power", cleaning up its atmosphere, restoring/maintaining the fertility of its farmland, and creating millions of worthwhile jobs (not just service-type gigs) and high-tech business opportunities for its citizens would be worth doing whether the reactors enabling it would cost $860 billion or three times that much. For comparison, Mark Jacobson, Stanford University's Pied Piper of the "green new deal" movement estimates that his 100% wind, water, and solar powered alternative (WWS) would cost us about $13.4 trillion paid for with big cuts in military spending, a very high carbon tax, and more highly taxed "wealthiest Americans" – not bond drives which would give the average US citizen a bigger personal stake in the outcome of any such campaign.

Of course, since I'm pretending here that the world's total human population will become 11.2 billion, the cost of the ~30 thousand full-sized reactors required to comfortably power everyone in it would be proportionately higher, ~30 (or 60) trillion of today's dollars. To put that number

[95] For example, during INL/Argonne's IFR project's last few years, its designers/modelers were told to deliberately reconfigure it to maximize "waste" burn up; i.e., waste as much precious startup fissile as possible. That's why Argonne's workers decided to switch the name of what they were working on from IFR to SFR ("sodium fast reactor"). Personally, I find it difficult to be cynical enough about what happens within DOE's nuclear complex.

into perspective, it's ~12% of what the "power walls" capable of backing up a similarly capable solar panel and/or windmill-based energy supply system for just one day would cost them.

"Special" materials that wouldn't be required by any of these nuclear reactors include the China-sourced rare earths required to make the relatively light-weight generators within the nacelles of most of today's wind turbines. There's no such weight requirement for any sort of ground-sited electrical generator, so nuclear power plants don't need them.

Green campaigners love wind turbines, but making their permanent magnets is environmentally impactful – to China, that is [148]. China currently meets about 96% of the world's demand for rare earths, and most of the so required mining, separation and extraction is done there. Every ton of rare earth element (REE) produced, apparently generates 9,600 to 12,000 cubic meters of dusty waste gases containing, hydrofluoric acid, sulfur dioxide, and sulfuric acid, ~75 cubic meters of acidic wastewater, and about one ton of relatively concentrated radioactive waste residue. China's Baotou region produces approximately ten million tons of such wastewater every year most of which is discharged without effective treatment and thereby contaminates both potable water and that used for irrigation. Disposal of the approximately 2000 tonnes of mildly radioactive mine tailings[96] generated per tonne REE is also problematic. Generally, tailings are placed into large open impoundments whereas in the U.S. permits are required and such waste piles must be properly covered or remediated in some other fashion.

A 2MW wind generator's permanent magnets require about 930 pounds (0.426 tonnes) of REE which means that building a wind farm with an average CF of 0.3 capable of providing as much energy per year as a 1 GWe MSR would require about 724 tonnes of rare earths - mostly neodymium [149].

To quantify this in terms of environmental impact, let's assume Hurst's contention that the mining/processing of one ton of rare earth minerals produces about one ton of fairly concentrated radwaste. In 2012, the U.S. added about 13.1 GW of wind power capacity which figure means that about 2770 tonnes of rare earths were used in the wind turbines installed here that year. It also suggests that about 2770 tonnes of the above-mentioned radioactive waste were created to make them.

To put that number into perspective, America's nuclear power industry produces about 2000 tonnes of spent LWR fuel "waste" each year. This means that the U.S. wind energy industry may very well have created more tonnes of radioactive waste that year (in somebody else's back yard, of

[96] Again, the reason why such stuff is somewhat radioactive is that REE and thorium are chemically similar and therefore usually present in the same ores.

course) than did its nuclear power industry. In that sense too, the USA's nuclear industry is relatively clean because it generated about 20% of its electrical energy that year while wind accounted for just 3.5 percent.

6.3 Waste Management Costs

> " *...eventually the antinuclear groups found the soft underbelly of the industry. It was something that had remained in the engineering background for decades. It was not nearly as exciting as striving for plutonium breeder reactor configurations or ceramic cores for jet engines, but it was there, and it was a distant bother. It was a nag, it was the long term disposal of all the radioactive byproducts of nuclear fission.*" ([150] p. 304)

Since the US is one of the world's most litigious countries, its reflexive anti nukes quickly realized that rendering waste management unnecessarily difficult would serve their purpose That plus the fact that DOE could be depended upon to spend lots of money "studying" it and then not accomplish much, turned reprocessing waste management into one of the biggest hurdles that a nuclear renaissance must get over.

The "should costs" of my scenario's waste management would be small because relatively little waste would be generated and it would be simple to first treat and then dispose of. Unlike today's fuel cycle, its waste would "burn" almost all of the especially troublesome long-lived transuranic radionuclides and thereby consist of only relatively short-lived fission products. The mass of such waste generated by 22 TW_e's worth of MSR-type power each year would be about 18,000 tonnes. Depending upon the type of reactor used, when everything required to isolate those fission products is combined with them and converted to glass, that glass would weigh from 15 to 40 tonnes/GW_e year and occupy 5-13 m^3 of space/year [151, 152]. These figures correspond to a total annual disposal volume ranging from 0.00015 to 0.00038 of one percent of that of the ash currently being generated by the world's coal-fired power plants and dumped somewhere each year[97]. Coal ash is mildly radioactive and contains a host of nominally toxic trace metals (e.g., mercury) all of which possess infinite half-lives, which

---

[97] Like mining wastes, coal ash often ends up behind huge earthen dams increasingly vulnerable to global warming-enhanced flooding. The resulting almost inevitable disasters routinely kill many people and pollute huge swaths of land downstream of those impoundments. http://theconversation.com/coal-ash-spill-highlights-key-role-of-environmental-regulations-in-disasters-1034 http://www.pollutionissues.com/Co-Ea/Disasters-Environmental-Mining-Accidents.html.

means that it represents a much greater (and proven) threat to both the environment and the people in it than does the waste generated by today's or the future's nuclear power plants.

6.3.1 Waste Treatment

Vitrification has been recognized to be the "Best Demonstrated Available Technology" (BDAT) for radwaste immobilization for about four decades. One of the reasons for this is that vitrification is intrinsically both simple and cheap which is why real world glasses typically cost under $3000 tonne - too cheap to be worth recycling in many places. Because the temperatures required to make them, are dangerously high (typ. 1000-1300°C), all glass making is done more or less "remotely" within thick-walled and thereby heavily "shielded" radiation-wise melters. Consequently, there's not much reason to assume that the cost of "remotely" making a radioactive glass should be vastly more than that, let's say $10,000 per tonne. Given that figure and a 15% mass-wise waste loading what would it cost to vitrify the ~18,000 tonnes of radionuclides generated/a producing 22.4 terawatts worth of breeder reactor power?

Tonnes of glass per year = 18000/0.15 =120,000 tonnes

Glass cost/a = $10,000/t*120,000 t/a=$1.2 billion/a

Vitrification cost per kWh = $1.2 billion/(22.4E+12 J/s*3600 s/hr*24 hr/day*365 days/a/3.6E+6 J/kWh) = $6.1E-6/kWh (trivial!!)

6.3.2 Waste Disposal

Disposal of such glass shouldn't be prohibitively expensive because one of the USGS's senior-most geologists [153], described a cheap way to accomplish it almost 40 years ago. Here's that paper's abstract:

> "*Portions of the Great Basin are undergoing crustal extension and have unsaturated zones as much as 600 meters thick. These areas contain multiple natural barriers capable of isolating solidified toxic wastes from the biosphere for tens of thousands to perhaps hundreds of thousands of years. An example of the potential utilization of such arid zone environments for toxic waste isolation is the burial of transuranic radioactive wastes at relatively shallow depths (15 to 100 meters) in Sedan Crater, Yucca Flat, Nevada. The volume of this man-made crater is several times (i.e., ~5 million m^3) that of the projected volume of such wastes to the year 2000. Disposal in Sedan Crater could be accomplished at a savings on the order of 0.5 billion, in comparison with current schemes for burial of such wastes in mined repositories at depths of 600 to 900 meters, and with an apparently equal likelihood of waste isolation from the biosphere.*"

That single, already radiologically contaminated/excavated, huge pit situated in the USA's most worthless desert could contain 20–50 years-worth of the future's HLW "waste forms".

The scientific basis of Dr. Winograd's concept was subsequently thoroughly investigated by SANDIA scientists, who dubbed it "Greater Confinement Disposal". It was subsequently implemented in the Nevada (bomb) Test Site's (NTS's) nearby"area 5" (APPENDIX X goes further into that story). A properly-sited GCD-type repository is key to solving the USA's interminable defense type radwaste boondoggling because it would eliminate the single greatest action-paralyzer/cost-driver in its approach to radwaste management – the ridiculously high volumetric costs assumed for such waste's burial plots which, in turn, serves to rationalize its decision makers' insistence upon trying to isolate such waste's highest stuff from everything else. Such huge scale, fully "remoted"[98] attempts at entropy-reversal are both unnecessary and prohibitively expensive.

Dr. Winograd's estimate (0.5 $billion) seriously underestimated the cost of DOE's ensuing radwaste burial schemes. It is also likely that he didn't suspect that DOE would end up charging US taxpayers roughly $1 million for each tonne (about 0.4 m^3) of "high level" glass that its contractors manage to produce[99]. The sole exception that I'm aware of was its one and only "Vitrification and Privatization Success" [154]. In that instance, a contractor (Duratek) promised to and then actually did convert 670,000 gallons of the Savannah River Site's liquid "mixed" (both radioactive and chemically toxic) radwaste plus some" hot" dirt and sludges to ~1700 tonnes of glass "gems" for $13.9 million (~$8400/tonne). Tougher-to-produce (higher melting) real-world bottle/window glass currently costs under $3000/tonne which renders it too cheap to be worth collecting/recycling in much of the USA.

---

[98] Almost all activities involving "hot" radioactive materials are performed "remotely", i.e, from a safe distance with shielding imposed between them and human operators.

[99] Again, in the DOE Complex, most of "vitrification's" cost is due to the separation activities (a.k.a., "pretreatment") preceding it in order to minimize the volume of the "high" fractions destined for its especially volumetrically challenged imaginary repositories. By August 2013, the Savannah River Site's glass melter had produced 3728 stainless steel canisters containing about 1.72 tonnes of glass each. That facility started up in 1996 and has received about $0.55 billion 2013 dollars per year since then. That works out to a glass cost of $1.46 million/tonne. Real world bulk glasses cost under $2000/tonne. http://srremediation.com/dwpf_40_canisters.html https://www.emcbc.doe.gov/SEB/SRSPostFY18Ops/Documents/Site%20Tours/2-SRS%20Industry%20Day%20-%20Overview%20-%20Spears%20-%20161027.pdf

"32N164W", Chapter 16 of Gwyneth Craven's "Power to Save the World" describes another technically competent, low cost waste disposal scenario thoroughly investigated by the world's scientists. GOOGLE EARTH reveals that it refers to a spot situated near the center of the Pacific Ocean far from the edges of any continental plate where the water is about twenty thousand feet deep and the basaltic bedrock underlying it covered with a ~thousand foot deep layer of soft pelagic mud/ooze. It's simultaneously about the most stable, most isolated, and least valuable piece of the earth's surface. Her chapter does a fine job of explaining why the World's top geoscientists concluded that it represents a truly excellent radwaste disposal site: big torpedo-shaped, steel encased glass (or concrete) radioactive waste forms dumped off a ship there would bury/spear themselves deeply within that mud thereby isolating their contents from the biosphere for millions (billions?) of years.

If such ships were to be powered with little THORCON or Terrestrial Energy IMSRs instead of diesel engines, transporting such radwaste forms to that disposal site wouldn't even be polluting the environment. The diesel fuel currently consumed by ships is a sulfurous, tar-like, "bunker oil" - not the relatively clean stuff sold by the first-world's service stations. It's literally the "bottom" of the barrel" both quality and impurity-wise which is why it's "suitable" only for an application that pollutes a "commons" (the oceans) rather than anyone's private property.

Incidentally, the world is already experiencing a steadily worsening "spent solar panel waste crisis". (To learn about this, see [155]).

It is unlikely that Goeller & Weinberg's cornucopian world could be implemented by either the world's poorest people themselves or the institutions that have traditionally provided most of their aid. Avoiding a deadly "hothouse earth" will require the simultaneous redirection of humanity's actions from selfish, for-profit exploitation to genuine stewardship and rapid transition to a genuinely sustainable economic system [156]. It won't be cheap or easy because rendering the world's energy systems GHG free, ensuring that its land, water and other resources are used sustainably, adapting to climate change and cleaning up an already polluted world will require changes that many influential people and the vested interests that they own/control will resist. The western world's national policies are largely shaped by the neoconservative belief that a market economy is the life force of civilization and that consumption is its purpose, thereby creating its jobs, wealth and material prosperity (Ayn Rand has become their leaderships' most-admired intellectual). That paradigm holds that producers will act for the common good guided by "sovereign" consumers without interference from government. By their thinking, if we just exhort individual consumers to purchase "green" products and services, we will eventually arrive at a clean/green/prosperous/fair form of capitalism. While there is some truth to that notion, it's inconsistent with the realities of corporate power and humanity's self-interest in general[100]. Corporations seek to maximize gains for their managers and shareholders and minimize environmental and other nonmarket obligations. That's why there's far more lawyers and lobbyists in Washington DC than elected "public servants". In a technological civilization containing millions of enthusiastic, intelligent, and eminently trainable young people, "saving the world" is really a political, not technical, problem and success will require the same extremely rare combination of technical smarts, unflinching commitment and hard-headed management skills exhibited by Admiral Rickover more than a half-century ago – not management exhibiting the symptoms of most of the USA's recent nuclear initiatives [157, 158].

Of course the problem is that DOE hasn't bothered to do the research necessary for anyone to decide what a Rickover-like program manager should build for us.

The following are the DOE nuclear project management "symptoms" identified by the National Academy's all time "best seller", "Barriers-to-science-technical-management-of-the-department-of-energy Environmental Remediation Program" [159] over two decades ago.

1. Planning that is driven by existing organizational structures rather than problems to be solved.
2. Commitments that are made without adequately considering technical feasibility, cost, and schedule.
3. An inability to look at more than one alternative at a time.
4. Priorities that are driven by narrow interpretations of regulations rather than the regulations' purpose of protecting public health and the environment.
5. The production of documents as an end in itself, rather than as a means to achieve a goal.
6. A lack of organizational coordination.
7. A "not-invented-here" syndrome at individual sites.

---

[100] In practice, those drivers translate to an economic system based upon "*let the buyer beware*" which also makes it difficult for buyers to determine what they're getting for their money (the technical details of the things we buy is often considered "proprietary" information)

An earlier report by the Galvin Commission [160], revealed a counter-productive federal system of operation" for DOE's national labs, saying, "*the current system of governance of these laboratories is broken and should be replaced with a bold alternative*". The problems it identified included, "*increased overhead cost, poor morale, and gross inefficiencies as a result of overly prescriptive Congressional management and excessive oversight by the Department,*" and "*inordinate internal focus at every level of these laboratories on compliance issues and questions of management processes, which takes a major toll on research performance.*"

Those outsider-generated reports had about as much effect upon DOE's subsequent behavior as does a spritz from child's squirt gun on a flock of ducks. Similar efforts by insiders (APPENDICES VII & VIII) didn't cause its decision makers to "embrace change" either.

6.4 How Do We Pay for It?

The most "obvious" source of the money required for the USA to develop and then implement a clean, green, nuclear powered future is its tremendously bloated defense budget. The majority of such spending is "discretionary" meaning that it could and should be used to address the energy-related technical issues affecting national security that a number of very senior military officers have noted [7].

Other obvious funding sources include:

- Revise the USA's tax structure to be more like it was back when America was genuinely "great" (late 1930's to circa 1970); i.e., heavily tax inherited wealth and raise the uppermost marginal income tax rate to at least 90%. Doing so would level the playing field for succeeding generations of US citizens and raise some (not most) of the money required for nation rebuilding. During 1944–45, "the most progressive tax years in U.S. history," the rate applied to any income above $200,000 ($2.4 million in 2009 dollars) was 94%. WWII's tax law revisions increased the number of "those paying some income taxes" from 7% of the U.S. population (1940) to 64% by 1944.
- "Green new deal bonds": The last time the United States issued what were then called war bonds was during WWII when full employment collided with rationing and they represented a good way to remove money from circulation

and reduce inflation. Issued by the U.S. Government, they were first called defense bonds, which name was switched to War Bonds after the Japanese attacked Pearl Harbor. Bond rallies were held throughout the country with celebrities to enhance advertising effectiveness. Free movie days were held in theaters nationwide with bond purchases serving admission and all proceeds deposited into the U.S. Treasury. In those days a median US yearly income was only about $2,000 despite which 85 million Americans — over half its population — purchased bonds paying about two thirds of that war's total cost.

- Carbon taxes

The USA's Manhattan project cost about $2 billion 1940 dollars (~$36 billion 2019 dollars). In just over three years, its workers turned a rather abstruse scientific observation[101] into massive technical infrastructure that succeeded in abruptly ending WWII and thereby likely saving a half million US lives. The USA's WWII bond drives raised over ninety times that much money, $185.7 billion 1940 dollars ($3.2 trillion 2019 dollars).

A recent evaluation of thirty-three studies looking at the return on investment (ROI) of infrastructure investments suggests that smart infrastructure development programs like the ones that China's been doing exhibit a 10-20% rate of return in terms of increased economic activity [161]. That suggests that the federal government's borrowing money from its citizens to first develop and then implement a "nuclear green new deal" would be a really great investment. I suspect that I'm not the only US citizen who would be delighted to have an opportunity to invest in something like that rather than in the mysterious "financial products" that my financial advisor has my savings invested in now.

---

[101] Over Christmas vacation in1938, physicist Lise Meitner and chemist Otto Frisch made a discovery that revolutionized nuclear physics and quickly lead to the atomic bomb; i.e., that when ^{235}U absorbs a neutron, it usually (about 80% of the time) splits into two smaller atoms (fission products) and 2 or 3 "new" neutrons.

7

The Nuclear Establishment's Self-Inflicted Wounds

Goeller and Weinberg's utopian future can't come to pass if those individuals most able to bear the cost of the necessary sweeping changes — those who've benefited the most from today's economic and political systems — don't help to pay for it. Since interfering with current business practices and taxing excessive wealth are anathema to most of the Western world's political leaders, it's likely that Eastern not Western counties will become the "outsiders" helping the world's disadvantaged people to implement this or any other scheme capable of achieving its goals. One reason for this is that most of the West's decision makers seem to believe that nuclear power's technical innovation should be left to the private sector, not government. While that might make sense for IT (hardware, software, and cell phone app development), it is unlikely to work for this purpose because reactor development is much "riskier" requiring a far greater investment of money, time, special materials, experimental work, and technical experts able to address a host of tough mechanical, physical, chemical, and, especially, legal issues - it can't be done by a few clever "hackers" in a rented garage. Its goals - achieving genuine sustainability, preventing further environmental damage, and providing a better life for everyone including the poor several decades off in the future - are also inconsistent with those of the majority of today's politicians and venture capitalists. Getting government "out of the way" might sound nice but won't work because every major energy breakthrough in recent history has received public support that moved an idea to proof-of-concept to demonstration after which the private sector heavily invested and thereby became richer [162].

Of course, the other reason is that many of the Western world's decision makers have apparently been sold on the notion that the future should/could be adequately powered by an "all of the above" mix of politically correct (non-nuclear), renewable technologies (see [124, 125], and, for opposing views, [66, 145, 163–165], and, have therefore supported/subsidized

Darryl Siemer (ed.) Nuclear Power: Policies, Practices, and the Future, (119–146)

efforts consistent with that paradigm. Also, if those decision makers weren't overtly hostile to anything "nuclear", they paid only lip service to the development of a sustainable nuclear fuel cycle. This is understandable because nuclear power's champions haven't presented a particularly compelling case and the majority of US politicians were originally lawyers and/or businessmen, not engineers or scientists, and therefore don't think quantitatively about technical issues – especially anything that's "controversial" and/or not apt to be too obvious to ignore until their political career has ended. Of course the people that work for them in the USA's more-or-less privatized national laboratory system (e.g., INL, ORNL, Hanford…) pretty much have to do whatever those politicians want because becoming a "poor team player" in that sort of working environment is tantamount to career suicide.

The most accurate measure of any country's commitment to accomplishing anything is how much money it is willing to spend trying. In the world's richest Western nation (USA ~ $20 trillion GDP) we hear things like, "The US Department of Energy (DOE) has selected projects to develop a pebble bed reactor and a molten chloride fast reactor to receive multi-year cost-share funding worth up to a total of $80 million" [166], and, that "Secretary of Energy Rick Perry Announces $60 Million for U.S. Industry Awards in Support of Advanced Nuclear Technology Development", which support is to be split between 13 projects in 10 different states [167]. Although much of this work will be performed in the government's laboratories by its contractors' employees (it's illegal to do genuinely "hot" research elsewhere in the US), its leadership won't assume responsibility for guiding such work or insist upon appropriate goals. Finally, although these projects have accomplished a good deal of computerized modeling, simulation, and "road mapping" (see Appendix IX and DOE's http://gain.inl.gov website), DOE's nuclear scientists/engineers can't effectively address the future's energy issues until they become empowered to do the same sorts of "risky" experimentation performed at its National Reactor Testing Station (NRTS), now INL, during the 1950's & 60's. Unfortunately, although those efforts successfully designed, built, operated, and then safely decommissioned ~50 different reactors, the USA never built a test reactor capable of evaluating today's front-running MSR concepts and is apparently still not planning to do so (see [168]).

In contrast to its "pretend" commitment to nuclear energy research and development, the USA's DOE/NNSA spends ~$6.9 billion/a to maintain its stockpile of nuclear weapons and its total (DOD+DOE) nuclear weapons-related expenditures are on the order of $20–40 billion/a [169]. "Nuclear modernization" is a euphemism covering a wide range of secretive activities that constitute in the view of many, a new, dangerous, and again, fabulously expensive global nuclear arms race. In the United States, the 30-year cost of the plethora of programs under its "nuclear modernization" umbrella – including new nuclear-capable bombers, land-based nuclear missiles, "mini bombs", and nuclear submarines – has recently been estimated at $1.2 to $1.7 trillion. Observers familiar with the Defense Department's $640 toilet seats suspect that if that entire modernization program were to be funded and carried out, its cost would be much higher even than that [170].

Here's a more up to date (10Jul2018) example of DOE's commitment to developing a sustainable nuclear cycle.

"U.S. Department of Energy Provides Nearly $20 Million for Domestic Advanced Nuclear Technology Projects" https://www.energy.gov/ne/articles/us-department-energy-provides-nearly-20-million-domestic-advanced-nuclear-technology.

Here's a breakdown of those awards.

$ million	Funded activity
7.00	Calendar Year 2018 Activities for Phase 2 of NuScale Small Modular Reactor project (a mini PWR)
0.498	Regulatory Support for Advanced Light Water Reactor Deployment: Advanced Boiling Water Reactor Source Term Reduction
6.314	Advancing and Commercializing Hybrid Laser Arc Welding (HLAW) for Nuclear Vessel Fabrication
2.101	Fluorination of Lithium Fluoride-Beryllium Fluoride (FLiBe) Molten Salt Processing – Flibe Energy
1.120	Experimental Verification of Post-Accident Integrated Pressurized Water Reactor (iPWR) Aerosol Behavior, Phase 3
1.925	Reactor Plant Cost Reduction to Compete with Natural Gas Fired Electrical Generation (Hitachi small modular BWR)
0.400	Conceptual Engineering for a Small Modular Reactor Power Plant Based on Lead-Bismuth Fast Reactor (LBFR) Technology

Finally, "DOE has selected two companies to receive GAIN technology development vouchers in this second review cycle. The companies selected are Yellowstone Energy (Knoxville, TN) in the amount of $160,000, and ThorCon US (Stevenson, WA) in the amount of $400,000. Further detail and description of these awards can be found under the GAIN website." (same reference)

Note that of DOE's $19.9 million commitment to "advanced" reactor development, only $ 2.5 million went to help develop anything having to do with a sustainable nuclear fuel cycle (Flibe Energy's fluorination demonstration and the Lead-Bismuth Fast Reactor's (LBFR) conceptual engineering).

"The President's Budget provides a total of $32.5 billion, $30.2 billion in discretionary funding and $2.3 billion in new mandatory funding in FY 2017 to support the Department of Energy in the areas of nuclear security, clean energy, environmental cleanup, climate change response, science and innovation". https://www.energy.gov/fy-2017-department-energy-budget-request-fact-sheet.

$2.5 million represents 0.0077% of DOE's budget and will be mostly spent supporting people working on "safe" (not risky) projects within its national laboratories.

James Mahaffey recently characterized the nuclear establishment's cultural norms as follows:

> "*Under fire nuclear engineering is to engineering as modern Islam is to religion. It's become more conservative and fundamentalist. Criticism from outside was so severe that internal criticism became more than the system could tolerate. The designs of nuclear reactors, plants, auxiliary mechanisms, and associated facilities, such as "waste" disposal systems or fuel handing strategies, all become more conservative*[102]*, with less engineering risk or innovation.*" ([150], p. XVI).

That culture manifests itself within DOE's nuclear complex as an over reliance upon authority, "procedures" (see APPENDICES VII, VIII, and XII), fine-sounding but irrelevant principles, secrecy, and a tendency to shade the truth a bit when dealing with outsiders (*if hard pressed, tell the truth but never the whole truth if there's any way it can be avoided* - see APPENDIX XII).

Government's role in this arena should be to fund long term, revolutionary projects, not sustaining and incrementally improving an already well performing LWR technology - that should be the domain of private industry. Government's job is to retire a <u>*worthwhile*</u> concept's risks so that private enterprise can then step in and capitalize upon that effort. Unfortunately,

---

[102] An American-type conservative embraces political beliefs characterized by respect for American traditions, republicanism, support for Judeo-Christian values, "business", anti-communism, individualism, advocacy of American exceptionalism, and defense of Western culture from the perceived threats posed by socialism, authoritarianism, and moral relativism.

Franklin D. Roosevelt said it better: "*A conservative is a man with two perfectly good legs who, however, has never learned to walk forward.*"

instead of restricting its scope to areas where academia and industry lack the funding, people, or facilities needed to innovate, DOE's nuclear engineering office (hereafter designated "NE") is engaged across the USA's entire nuclear enterprise, spreading its focus and expenditures over myriad, disparate, & mostly not cutting-edge, activities. Evidence of its lack of focus is the blizzard of road mapping and strategy documents prepared by its employees. The thirty anonymous experts interviewed by Ford *et al.* [158] characterized NE's activities as follows:

1. "*yes, we have enough roadmaps to publish an atlas. And yet, no vision.*"
2. NE's real goal is to maintain its funding stream, "*flying under the radar to the greatest extent it can in order to avoid political controversy, and it generally succeeds at that.*"
3. NE's project funding is an "*old boys' club,*" where investigators are funded, "*if NE had funded them in the past.*"
4. NE favors funding "*known quantities*" in order to "*prevent surprises.*" Evidence of good performance or innovative research rarely comes into the equation[103].
5. NE is most definitely not interested in "*taking risks:*" it neither rewards nor encourages radical deviations from its programming norm.

Those experts lamented the fact that the U.S.A.'s nuclear R&D enterprise is led by an organization that avoids taking risks and making hard decisions, frowns upon ambitious, long-term projects, funds them at a low level and is primarily concerned with the next appropriations cycle.

More than two-thirds of them opined that NE's national laboratories ought to be mainly a facilitator, or enabler, of research. They should conduct high-risk and potentially high-reward research and maintain the facilities that buttress industrial innovation, as opposed to micro-managing its activities.

7.1 Refusal to Choose/Set Rational Goals

There's been a great deal of confusion about what a nuclear renaissance should do and even more about what its implementation issues would be.

---

[103] Again, within NE, a project's products are reports (paper) and, occasionally, a decision – nothing concrete.

In spite of the fact that almost everyone working for DOE knows that we live in a world that will soon need far more "new" clean energy than Jacobsen *et al.*'s 100% wind, water, and solar power scheme could supply to world containing >11 billion people, its decision makers refuse to admit that that's the primary reason for nuclear engineering/scientific research and development. Consequently they encourage (force) the people that work for them to devote their efforts to developing reactors/fuel cycles suited for addressing short term political needs; i.e., "less proliferative", "smaller", "modular", "idiot proof", "zero risk", "waste burning", "simpler", etc.[104], not saving the world. It's unreasonable to expect individuals working within any such top-down driven system to devote much attention to developing anything capable of providing 100% of our grandchildren's energy.

Let's take a look at some examples.

7.1.1 NGNP

A recent such example was DOE NE's Next Generation Nuclear Power (NGNP) boondoggle intermittently funded from 2005 to 2013. Its champions invoked huge, high pressure, helium cooled, graphite moderated, solid-fueled, "very high temperature reactors" in order to generate hydrogen *somewhat* more efficiently (i.e., chemically) than does water electrolysis[105]. Of course, doing that would require that its new high temperature gas cooled reactor (HTGR) operate at very much higher temperatures (~1000°C) than had any of the other gas-cooled reactors previously built/tested, meaning that lots of time and money would have to be spent developing "special materials". HTGR represents an old, much investigated, and commercially

[104] For instance, during most the last two decades, it's been verboten to use the word "breeder" in any report describing reactor concepts that DOE was considering. That's like telling NASA's engineers that they can no longer mention (or consider) "rockets".

[105] During this period, circa 2000-2013, DOE's vision for the future invoked a "hydrogen economy" in which most of the things currently powered with petroleum would be powered with hydrogen instead. That never made much sense because hydrogen is an extremely volumetrically challenged energy carrier (not fuel) – at normal atmospheric pressure, burning it generates 0.00034 as much heat as does burning same volume of petroleum. While fuel cells are 2-3 times more energy efficient than are hydrogen-burning heat engines, it's still very difficult to imagine how cars and trucks could be safely powered with hydrogen due to its uniquely low mass and boiling point - tanks strong enough to carry the equivalent of 10 gallons of gasoline would be impractically big/heavy. Second, if we really wished to make more hydrogen, the goal should be to generate electricity more sustainably and cheaply, not to come up with yet another "advanced" converter-type (unsustainable) reactor concept that could only be built of unobtanium.

unsuccessful reactor concept[106]. It turned out that such material would be extremely expensive (or outright impossible) and, since any such fuel would be almost impossible to "reprocess", it would also be almost impossible to render any such system genuinely sustainable. Consequently, DOE's best and brightest nuclear scientists and engineers wasted another decade studying something that the USA's civilian nuclear industry wasn't interested in and doesn't represent a practical solution to the world's long term energy problems.

The NGNP program's other issues were largely due to NE's lack of programmatic discipline. It tends to micromanage its grants to an intrusive extent; it rarely follows through on any advanced, non-light water reactor project; it does not fund them at the level or duration necessary for project success; and, it is so attuned to political sensitivities that it often abandons just-started programs in favor of others deemed more politically palatable. Some factors beyond NE's control such as the inflexible cost-sharing arrangements mandated by Congress and the Office of Management and Budget (OMB) also make it difficult for industry to collaborate with it.

Of the approximately 400 papers and posters presented at the IAEA's latest international conference having to do with "*Fast Reactors and Related Fuel Cycles: Next Generation Nuclear Systems for Sustainable Development* (FR17)" [171], four presentations were made by people representing the USA – one was about DOE's latest road mapping efforts, two about its long-defunct LMFBR program, and one by an antinuke anxious to remind everyone that anything nuclear could be dangerous.

Another reason for the USA's poor showing at such international gatherings is that its current regulations criminalize many of the activities intrinsic to international cooperation[107]. Bill Gates recently summed up that situation

---

[106] The last "serious" attempt to get a HTGR up and running ran from 1993 to 2010 in South Africa. It would have been the first Generation IV unit to enter the construction phase (but didn't). Its pebble bed (PBMR) design was based on a German technology demonstrated during the 1970s and 80s. It was to be graphite moderated and helium cooled with a closed-cycle gas turbine based on the Brayton thermodynamic cycle. It had a proposed thermal output of 400MWt and electrical output (net) of 160 MWe. Its fuel would comprise thousands of ~6 cm diameter spherical graphite "pebbles", each filled with thousands of tiny silicon carbide and pyrolytic carbon coated UO_2 particle, tristructural isotropic (TRISO) fuel kernels which were supposed to retain all fission products.Its fuel pebbles had a life expectancy of three years and were to be circulated through the core about six times before replacement. DOE's NGNP project designers assumed that their new & improved TRISO kernels would be embedded in big prismatic graphite logs fixed within the core, not circulated, until their fissile had burned up.

[107] For instance, any research finding that the DOE's bureaucrats deem potentially useful to "enemies" is classified and thereby rendered unavailable to anyone else working on the same problem. One of the most absurd manifestations of that mindset was that during

as follows, "*Unfortunately, America is no longer the global leader on nuclear energy that it was 50 years ago. To regain this position, it will need to commit new funding, update regulations, and show investors that it's serious*" [172].

Those cultural characteristics along with a number of embarrassing multi-billion dollar boondoggles have led many outsiders to be leery of anything having to do with nuclear power.

In a finite world, money wasted doing unnecessary things is money that can't be spent doing necessary things. Let's take a closer look at some of those unnecessary expenditures.

7.1.2 DOE's Savannah River Site's MOX Boondoggling

After considerable study during in the early 2000's, DOE concluded that the best/cheapest way to "manage"[108] the US federal government's ~34 tonnes of surplus weapons/bomb-grade plutonium would be to fabricate MOX (mixed oxides of plutonium and uranium)-type fuel for US civilian power plants[109]. This isn't a novel idea or even a particularly difficult thing to do because the French, British, Russians, and Japanese had already been making MOX of the far more radioactive, poorer quality (lesser fraction of ^{239}Pu) "reactor grade plutonium" recovered by their fuel reprocessing operations. Such fuel's proposed US customers didn't really want it because it's quite "hot" and would therefore likely cost them more in terms of increased training and regulatory overhead than it would be worth[110]. Actual construction started

---

the early 1980's, someone blanked-out the symbol for ------(-) on our lab wall-mounted periodic table because it was the "secret" element rendering the USA's nuclear navy's metallic zirconium/uranium fuel assemblies extra special. I guess the rationale was that any Russian spy who broke into our lab wouldn't be able to figure out what that mark was covering. The modern periodic table was invented by a Russian, Dmitri Mendeleev, 150 years ago and the USA doesn't possess secrets about reactors that the Russians and Chinese don't already know (if anything it's probably now the other way around).

108 In DOE-speak, "waste management" might mean anything from the "no action alternative" (doing nothing which in cases like INL's already calcined wastes or SRS's "excess" bomb grade Pu, is likely the best option) to promising to transmute whatever it is to diamonds and ship them to the other side of the moon. It shouldn't surprise anyone that such exercises' almost inevitable "mission creep" tends to be in reverse. In the DOE contracting business, the biggest liar usually wins, which means that DOE does lots of goal post shifting to cover its mistakes.

109 Which decision apparently ignored Bolgren *et al.*'s conclusion that such a facility would cost about $50 billion [94].

110 In today's uranium market, the raw uranium going into a light water power reactor's fuel assemblies represents only about 25% of their cost which, in turn, represents only 20-25% of the reactor's total operational cost, most of which is the "overhead" required

in 2007 but DOE's management program "symptoms" [159] combined with the political pull of So Carolina's politicians quickly turned that project into another profitable (for them) boondoggle. In 2007, DOE's experts promised that its MOX plant would cost $4.8 billion and be completed by 2016 [173]. As of a year ago, assuming flat $350 million per year continued funding, they estimated that it would cost $17.2 billion and take until 2048 to complete (this sounds a lot like DOE's predictions/promises having to do with its Hanford site's waste "vitrification" project)[111]. In 2014, the Energy Department finally admitted that plutonium could be disposed of far more cheaply using a clever idea labeled "dilute and dispose" (patented?) but that notion was opposed by South Carolina's political leadership headed by the redoubtable Lindsey Graham because it might kill what had become a very fat and dependable golden goose to his constituents. Consequently, from 2014 to 2016, Congress repeatedly gave DOE the same message: "*keep building the MOX plant.*" In 2017, Congress authorized energy secretary Perry to stop construction if he could show that another approach would cost under half that much. In May of 2018, DOE promised that if it were allowed to stop, that building's shell could be converted to an equally or maybe even fatter "manufacturing plant for nuclear weapons" - type goose. As of May 2018, DOE was spending ~$1.2 million per day studying its options [174].

That batch of bomb grade plutonium isn't "waste" because it cost US taxpayers many $billions to make and could serve as the start-up fuel for 4-15 full-sized breeder reactors[112]. This means that we should continue to store it until it has been determined for sure which breeder concept represents the best way to implement a nuclear renaissance.

7.1.3 DOE Hanford's Reprocessing Waste Treatment Project's Boondoggling

Some time ago another QUORA's reader asked me, "*Why do we continue to develop more nuclear energy and products when we cannot safely get rid of the radioactive waste*?"

to support the people running/maintaining/overseeing it. Again, for such fuel about 80% of the raw uranium mined is discarded during its enrichment step and never ends up in a fuel assembly. Such "depleted" uranium is currently considered waste although a fraction of it is utilized to make military tank armor and ammunition.

111 At $17.2 billion, such fuel would be well over twice as costly as are today's "mined uranium" 235U-based fuel assemblies.

112 Rumor has it that that at least some of it is now being turned into fresh "pits" for the USA's bomb inventory. Due to its multiplicity of allotropes, metallic plutonium tends to "age" meaning that such pits occasionally need maintenance or replacement.

Like many of QUORA's questions this one is somewhat misleading because the US is already implementing a perfectly reasonable way of dealing with that sort of "waste" – physically intact "spent" raw fuel assemblies generated by its civilian power reactors. That's reasonable because, 1) it's both safer and cheaper than is first reprocessing those assemblies and then dealing with the various radioactive waste streams sogenerated; and, 2) the MOX-type LWR fuel made from recycled plutonium isn't worth what it costs to make. However, if the future's decision makers were to decide to address mankind's environmental and economic challenges with an appropriately scaled "nuclear renaissance", they/we would have to switch to a closed nuclear fuel cycle capable of efficiently burning such stuff and resume spent fuel reprocessing (actinide recycling). Today's set of self-defeating rules, laws, and assumptions render that rosy scenario impossible because they prevent the USA's technical people from dealing with its already existing reprocessing wastes in a rational fashion which is, of course, exactly what the USA's anti-nukes love to see – radwaste management has become a "transcientific", not technical, problem. Unfortunately, it also provides the jobs supporting lots of technical people, their management and many

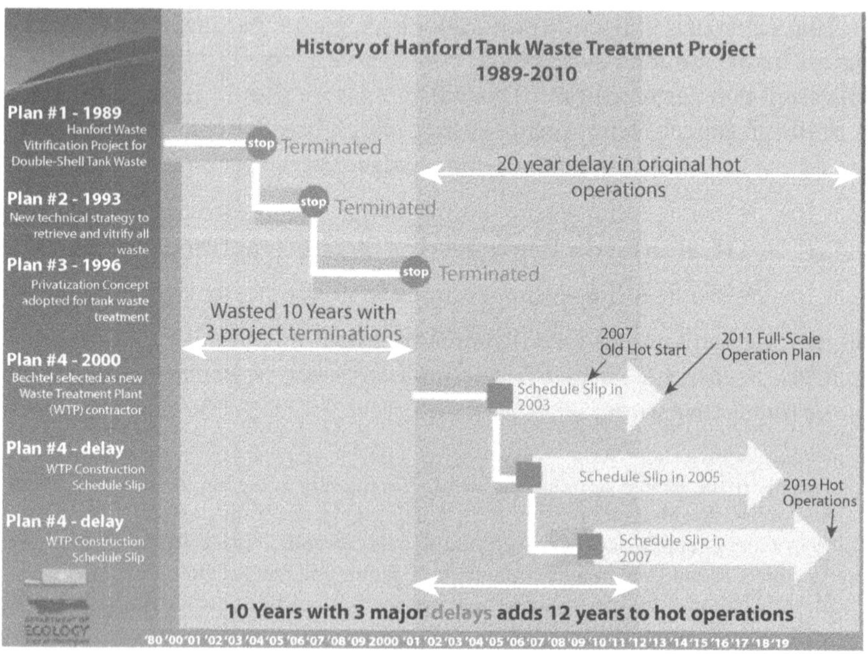

Figure 24 History of the Hanford WTP 1989-2010 (source: Suzanne Dahl WA ECOLOGY, 2010).

"outside" advisors/helpers/officials/critics, meaning that almost no one questions the paradigm that's been driving Hanford's radwaste boondoggling for over three decades.

The root cause of DOE Hanford's Waste Treatment Plant's (WTP) interminable cost overruns, delays, etc. (see Fig. 24)[113] , is that politically correct but technically unrealistic assumptions morphed into "promises" nearly impossible to keep. Its key technical assumptions that Hanford's tank wastes will be separated into "high" and "low" fractions so that the former can be dumped somewhere else and that both fractions will be converted to a borosilicate-type glass – don't make sense for the following reasons:

- The ~55 million gallons of salt-wastes in Hanford's tanks is extremely dilute radionuclide-wise and now old enough to generate relatively little heat which means that it is "high level" only due to its origin/history/associations[114] (see APPENDIX IX)
- The probability that another US state would ever choose to host a repository for the "highest" stuff that might eventually be isolated from Hanford's (Washington's) waste tanks is low
- Hanford's tank waste is situated well above the water table in one of the USA's driest deserts (it's in a perfectly adequate but not "perfect" repository site)
- That site is already heavily contaminated – DOE's hypothesized destination for Hanford's "highest" stuff remains pristine almost four decades after ~15 billion taxdollars were wasted "studying" it.
- The huge-scale radioisotope separations (aka "pretreatment") required to isolate its waste's especially "high" fractions are intrinsically difficult – mixing such stuff is easy/cheap, separating it (entropy reversal) is tough/expensive
- Pretreatment requires additional chemicals which would inevitably create still more radwaste
- DOE's one and only "Vitrification and Privatization Success" [154] didn't invoke pretreatment – everything in those

[113] Figure 24 was excerpted from slide set presented by a Washington state official nine years ago describing what had happened during the first 21 years of that project. At that time, Hanford was working on implementing its fourth official plan - it's now working under its eighth official plan.

[114] The AEC's approach to fuel reprocessing stressed "productivity" and did not prioritize the minimization of ash forming additives (metal containing) which is the reason that the bulk of DOE's HLW consists of non-radioactive sodium, potassium, aluminum, iron, etc., salts.

particular radwaste tanks was simply mixed with powdered sand & borax and then vitrified (melted) to produce borosilicate-type glass "gems" (irregular obsidian pebble-like marbles)
- Hanford's tank wastes contain large amounts of non-radioactive materials (e.g., sulfate and the halides) incompatible with the sort of glass (borosilicate) that its decision makers insist upon
- More suitable glasses have been extensively researched in the USA [175] and extensively employed elsewhere [176].
- Such glasses would be more chemically compatible with cementitious "grouts" (pumpable slurries that set up to form concrete) than is borosilicate glass.

Consequently, a more reasonable approach would be to mix, not separate, Hanford's tanked radwastes and vitrify that mixture to make iron phosphate glass "gems" which would then serve as the aggregate of a cementitious grout pumped back into those same tanks for disposal.

Here's why this proposal makes good sense:

- Hanford's tanks (Fig. 18) are neither evil nor apt to be going anywhere.
- They have already been paid for – the ~106,000 giant "transportable" (to where?) stainless steel canisters that the DOE's current plans call for have not.
- Hanford's much larger (typ, 1 million gallon), steel-lined, reinforced concrete, underground "canisters" (tanks) (Figure 25) are apt to more durable over the eons than its current plan's plastic-lined "not-high" radwaste disposal pits would be.
- Doing this would simultaneously remediate those tanks themselves - filling them with grout (concrete) would immobilize any residual waste, seal them, and render them both crush and corrosion resistant.
- It would obviate the current plan's huge "interim" "high level" waste form packaging, storage, and (hypothesized) retrieval, repackaging, offsite transportation and disposal costs.
- It would be safer than the current plan because it's a great deal simpler, meaning that "operators" wouldn't have to do nearly as many things with/to that waste.

Consequently, it would also be much, much, cheaper.

Implementing my scheme with Hanford's vitrified/grouted tank wastes would utilize under 30% of the space within its tanks because the glass

Figure 25 Hanford's waste tanks during construction (typically 1 million gallon each)

would be more compact than the raw waste going into it[115]. Since the Hanford site currently possesses many other sorts of radwastes ("crib" dirt, sludges, etc.) for which no solution has yet been implemented, most of them could and probably should be converted to the same sort of durable "aggregate" and grouted into those same tanks too. When that has been accomplished, it would then be reasonable to fill any remaining head-space with grout serving double-duty as a waste encapsulant for low-level "orphan" wastes[116] imported from elsewhere.

When this scheme was presented to Hanford's decision makers and their advisors six years ago [288], I pointed out that they could probably levy substantial fees for providing such a noble service for other states and National Laboratories. I also mentioned that Congress may eventually decide to quit funding their giant boondoggle (~$59 billion at that time) which would eventually downsize everyone employed by it.

Of course, Hanford's decision makers ignored me because they had already made too many silly assumptions, pronouncements, and promises based upon what the US federal government's radioactive waste management experts had told them ~30 years before, and, more importantly, no sufficiently influential "outsider" was forcing them to shift their paradigm. That's why that project still hasn't accomplished much other than continue to

115 The reason for this is that such waste's major components, both mass and volume wise, are thermally labile species such as nitrate, nitrite, and water which would not end up in any glass.

116 "orphan wastes" don't have a designated real or imaginary destination (home) yet.

provide ~$2.5 billion/year worth of "work" for the horde of technical, legal, and bean counting experts struggling with that paradigm's multiple issues.

That project's leadership and local stakeholders are apparently getting even more greedy/ambitious: just today (1Feb2019), the Hanford site's local newspaper reported that,: "*The increase in costs for the tank waste treatment and disposition increases from an estimate of $53.5 billion in 2016 to a range of $221.4 billion to $518.1 billion in the report released Friday*" [177].

The USA's electorate's docile acceptance of such self-serving nonsense from the people who nominally work for them is what's turned most of their governments' reprocessing waste management exercises into long festering boondoggles - Hanford's just happens to be the biggest example of it.

Change may finally be at hand: On 18Mar2019, the Trump administration responded to Hanford's request for still more money by cutting its budget by $416 million (from $2.5 to 2.1 billion) https://www.seattletimes.com/seattle-news/trump-administration-proposes-big-cut-in-hanford-spending/.

7.1.4 DOE's "Lead Nuclear Engineering Lab's" Radwaste Boondoggling

The AEC/ERDA/DOE's NRTS in Idaho had its own little fuel reprocessing facility to recover the highly-enriched uranium utilized in such testing so that it could be reused (about 50 different reactors were pilot-planted there from circa 1950 to the early 1970's). ~90% of the liquid processing waste thusly generated was converted to a relatively stable/safe mix of dry dust and sand-like granules ("calcine") which was and still is, stored within stainless steel tanks within reinforced concrete vaults ("binsets") When President Carter decided to end the USA's nuclear fuel reprocessing activities, US DOE's nuclear reactor R&D pretty much ground to a halt too, meaning that a "new mission" had to be found for its Idaho Site[117].

---

[117] Its most recent "old mission" had been to recover the highly enriched uranium in spent naval reactor fuel. INEL's "Chem Plant" had been almost totally refurbished to do so by the time that I "came on board" in 1978 – well over a billion 1970s-type dollars had been spent to build a new fuel dissolution system, fuel storage facility, and New Waste Calcination Facility (NWCF). Unfortunately, its dissolvers were designed to dissolve conventional zirconium (or "Zircalloy") clad fuel assemblies – not the Navy's "special" zirconium/uranium fuel assemblies alloyed (coated?) with the secret element that no one was allowed to mention. That fact was apparently kept a secret from that facility's designers. Anyway, it didn't work very well and the Navy eventually decided to quit pounding money down DOE's rathole. Uranium recycling is/was irrelevant to naval operations – it's got cost-plus fuel contracts and the recovered uranium was not going to be reused in its own reactors anyway (it's too radiactive).

Therefore, it eventually became DOE's "lead laboratory in radioactive waste management" and was renamed the "Idaho National Environmental and Engineering Laboratory" (INEEL). Because the other DOE laboratories which had been running reprocessing facilities (Savannah River and Hanford) had already decided how their wastes were to be treated, INEEL's new mission really just boiled down to devising "uniquely cost effective ways" of dealing with its own. The first priority would be its remaining ~1 million gallons of still-liquid "sodium bearing waste" (SBW) because it's less stable and more mobile and therefore represents a greater threat to the good folks of Idaho than does its already-calcined stuff. Since INEEL's calciner was still working at that time (1995), it would have been reasonable to simply add some sugar[118] to its SBW, calcine it, and store it in its already built and paid for stainless steel/reinforced concrete "bin set number 7" until someone decides for sure what's going to be done with INL's calcines. That process had been developed and successfully pilot plant tested many times in the past. However, the folks who had previously managed the Idaho site's reprocessing facility had become the leaders of its new waste management mission too, which meant that it was to be approached in the same fashion; i.e., calcines would be retrieved from their bins and dissolved in nitric acid so that some of their especially "high" radionuclides could be isolated and converted to hot isostatically pressed (HIPed) super-ceramic waste forms. In other words, an all-new facility would be built to "reprocess" calcine instead of spent reactor fuel and then convert the highest fractions so isolated into super durable, theoretically dense (maximally compact), radioactive waste forms. The rationale dreamed up for that brain storm was that, "*the most problematic/expensive characteristic of INEEL's anticipated reprocessing waste forms would be their physical volumes*"; i.e., that there wouldn't be enough "space" in DOE's Yucca Mountain repository to accommodate INEEL's reprocessing radwaste if its volume weren't somehow reduced first. Of course, that really doesn't make much sense because INEEL's ~4500 m^3 of calcine could certainly have been fit into the ~370,000 m^3 of space (fives miles worth of 25 ft-diameter tunnels) that DOE's YM study site's tunnel

[118] Sugar serves as a chemical reductant for nitrate (or nitrite) and provides the carbon dioxide required to convert the resulting alkali metal oxides to relatively high-melting carbonates. This simultaneously prevents "bed agglomeration" and greatly reduces the amount of gaseous/toxic NOx generated (most of the nitrogen ends up as harmless N_2 instead). British and French reprocessing facilities sugar-calcine the wastes going into their HLW glass melters with close-coupled rotary kilns.

boring machine had already created[119]. However, DOE's decision makers really liked that scenario because it was extremely complicated, meaning that lots of research would have to be done to work out its implementation details, which translated to job security for its field office personnel and their contractors' favorite experts and managers. DOE's decision makers would also be able to put off "final" decision-making (their job) until all of that research had been performed, properly reviewed, revised, and documented[120]. Idaho's most important stake holders also liked it because another couple decades worth of federal funding would be coming in to fund that/their "terribly important" project.

Well, since SBW is/was already a nitric acid/nitrate salt solution containing some especially "high" components (TRU (transuranic elements), ^{90}Sr, ^{137}Cs, and ^{99}Tc, it *seemed* to make sense to not calcine it as had been promised [179] but instead use it to test the as-yet undeveloped separation technologies that were to be applied to its hypothetically redissolved calcines (this is a typical example of what DOE's "road mapping" exercises accomplish). Consequently, its remaining liquid reprocessing waste never was calcined. When INEEL's "outside" stakeholders finally realized that its separations-based waste management scheme didn't make sense, its "waste side" contractor had had already permanently shut down its calciner[121] and

---

[119] It was also inconsistent with the statute defining that hypothetical repository's "capacity"; i.e., stuff containing radionuclides equivalent to that within 70,000 metric tons of spent ~33GWd/t power reactor fuel, not 70,000 (or any other number of) "cubic meters". An independent evaluation of INL's accumulation of calcines, spent fuel, etc. performed several years earlier had concluded that its HLW totaled up to an equivalent of about 320 tonnes of spent power reactor fuel [178]. To make its brain storm appear more rational to its stakeholders, DOE decided to claim that each cubic meter of INL's calcined radwaste is equivalent to 0.5 (not 320/4500) m^3 of spent fuel and that the price that it'd be charging US taxpayers to store high stuff within its already bored-out YM repository would be ~\$800,000/m^3.

[120] Everyone working for DOE in one way or another (civil service, contractor/subcontractor personnel, and the academics it chooses to fund) is rewarded for doing/saying things consistent with its current thinking. Consequently, during those times when its leadership apparently isn't interested in actually solving problems (e.g., while INEL/INEEL was DOE's lead lab in radioactive waste management), coming up with a fresh excuse for not solving them is career enhancing.

[121] The rationale ginned up for shutting down INEEL's calciner (NWCF) was that because it was a waste "incinerator" it would have to be rendered MACT compatible (i.e., modified so that it not emit a huge cloud of NOx) which, in turn, would "*cost taxpayers too much*" (~\$50 million). What wasn't mentioned were that; 1) calcination isn't incineration because the waste in question isn't flammable, 2) sugar calcination would have greatly reduced NOx emissions, and 3) calcination could have been completed well before the EPA's MACT compliance deadline (June 19, 2001, see [180]).

the deadline to complete the "treatment" of its SBW was now only about 11 years off (i.e., to be completed by 31Dec2012 – the deadline for completing calcine treatment/disposal was and is still 31Dec2035). This sparked a renewal of the waste vitrification (glass-making) research that had been limping along during most of the previous three decades [181]. However, that option also vanished when in late 2002, DOE's brand new "EM-1" (Jesse Roberson) declared that vitrification would be "too expensive" for INEEL (but still OK for both Savannah River and Hanford – she'd apparently already been traumatized by the latter's "vit plant" boondoggling). Consequently, INEEL's decision makers had to look for something else that would be uniquely cost effective. Ignoring several "outsiders"[122] and one insider's advice (mine –[183]), they settled upon "steam reformation" because it invoked a mysterious ("proprietary" therefore perfectly OK for both DOE and its contractors to keep secret) but apparently successfully patented, fluidized bed-based "steam reforming" technology[123] rather similar to the calcination that INEEL's managers and engineers were already comfortable with. Other pluses were that the subcontractor in question claimed that its process could make a variety of products suitable for any sort of repository anywhere, and that it would only cost about $45 million to treat INEEL's SBW.

APPENDIX IVX describes the steam reforming process and explains how it came to be DOE's "preferred alternative".

This story is already getting too long, but suffice it to say that: 1) circa 2006, DOE decided that steam reforming's easiest-to-make (and least durable; i.e., water soluble) "carbonate product" would be good enough for its not-so-high-level repository site[124]; 2) all attempts to get even that

---

[122] "*DOE-ID should not pursue further steam-reforming initiatives for treatment of SBW to produce waste forms for direct disposal in a federal HLW repository or in WIPP*" [182].

[123] Steam doesn't "reform" anything other than some of the solid carbon (instead of dissolved sugar) serving as the reductant in that sub contractor's (THOR's) proprietary technology (i.e., $C+H_2O \rightarrow CO+CO_2+CH_4+H_2$ etc.). Unfortunately, much of that reductant (generally coal) is simply blown up and out of the >>600°C reactor along with the other "fines" which collectively constitute the bulk of its product. In principle, but not in practice (too problematic), it is possible to convert SBW's ash-forming elements (mostly sodium, potassium & aluminum) to a more durable (not as water soluble) nepheline-like, alkali aluminosilicate mineral assemblage by adding some clay too [183]. See APPENDIX IV.

[124] i.e., the Waste Isolation Pilot Plant (WIPP). That decision was likely due to the fact that none of the "demonstrations" of the sub contractor's (THOR's) attempts to make the superior (not so water soluble) product specified for DOE's HLW (YM) repository had worked as had been repeatedly claimed. That information was only released to me via a freedom of information request [184] – up until then, the "raw" data/reports generated by its subcontractor's (Hazen's) tests/demonstrations had been kept under wraps due to their "proprietary" nature.

relatively simple process actually running have since failed; 3) 100% of INL's liquid reprocessing waste remains in its tanks over six years past DOE's promised 31Dec2012 treatment and disposal deadline; and, 4) INL's still-continuing steam reforming boondoggling has probably already cost US taxpayers well over a billion dollars (more exact cost figures are unavailable to "outsiders").

In the DOE Complex, "troublemakers" are ignored because there is no reason not to: programmatic waste, fraud, abuse and failure has become the norm for its nuclear-related projects and the only thing that really counts with regard to individual personal career development is being an organized good team player[125]. Also, in every case I've seen & tested[126] (several times), the outside stakeholders that really count (politicians) aren't much interested in anything beyond maintaining local spending & employment figures[127].

The worst news is that circa 2005 INEEL was reacronymed "INL" and its leadership given a new and much more important mission to manage – it became DOE's "lead nuclear engineering laboratory" and has remained so to the present.

Its NE R&D projects have failed to accomplish its new mission for the same reason that its signature technical waste management project (SBW "treatment") did – its managers neither motivate nor empower its scientists and engineers to do what's necessary to succeed.

I'm not alone in having that opinion. "*A Retrospective Analysis of Funding and Focus in US Advanced Fission Innovation*" jointly funded by

---

125 Another reason is that DOE's program managers don't take responsibility for any sort of technical goof and are rarely punished for failure (contractors get the blame). The same applies to the management of the institutions nominally responsible for overseeing its activities. In the DOE Complex, no one seems to be responsible for anything other than complying with rules and procedures. If program oversight doesn't mean accountability, it doesn't mean anything. That's totally counter to Rickover's management philosophy: "*You may delegate it, but it is still with you... If responsibility is rightfully yours, no evasion, or ignorance or passing the blame can shift the burden to someone else. Unless you can point your finger at the man who is responsible when something goes wrong, then you have never had anyone really responsible.*"

126 I have yet to hear back from any of the politicians & officials to whom I've sent letters (e.g., APPENDICES VII & VIII).

127 The fact that DOE's locally important political stakeholders routinely agree to project completion dates ridiculously far off in the future supports this probably "controversial" contention. If in 1995, Idaho's Governor Batt had given DOE, let's say, five years to complete the calcination of INEEL's reprocessing wastes, that's what would have happened. Allowing DOE 17 years to accomplish something that could/should have easily been done within two years effectively issued it & its contractors licenses to steal.

the John D. and Catherine T. MacArthur Foundation and six other agencies independent of the US DOE [157] concluded that, "*Using extensive data*[128] *acquired through the Freedom of Information Act, we reconstruct the budget history of the Department of Energy's program to develop advanced, non-light water nuclear reactors. Our analysis shows that—despite spending $2 billion since the late 1990s—no advanced design is ready for deployment. Even if the program had been well designed, it still would have been insufficient to demonstrate even one non-light water technology. It has violated much of the wisdom about the effective execution of innovative programs: annual funding varies fourfold, priorities are ephemeral, incumbent technologies and fuels are prized over innovation, and infrastructure spending consumes half the budget. Absent substantial changes, the possibility of US-designed advanced reactors playing a role in decarbonization by mid-century is low.*

Another paper published that same year, "*Expert Assessments of the State of U.S. Advanced Fission Innovation*" [158] concluded that "*.... results from structured interviews conducted with 30 nuclear energy veterans to elicit their impressions of the state of U.S. fission innovation. Most experts assessed NE as having been largely unsuccessful in enabling the development of advanced designs. The interview results highlight the importance of leadership and programmatic discipline, and how their absence leads to poor performance in driving change. Responses point to the likely demise of nuclear power and nuclear science in the U.S. without significant improvements in leadership, focus and political support.*"

Amen.

7.2 Fukushima's "Nuclear Disaster"

In 2011, an eminently predictable earthquake and subsequent tsunami off the east coast of Japan did about $300 billion worth of property damage and killed about 9,000 people. Nevertheless, about the only thing that we hear about is the Fukushima Daiichi nuclear power plant's "nuclear disaster" which killed no one[129] and damaged only that facility's equipment and buildings. Several days after its reactors had been safely shut down, the cooling water in three of them and some adjoining fuel assembly storage

---

[128] Abdulla *et al.* had to sort through a 400,000 page FOIA data dump to come up with the numbers leading to their conclusions [157] – read it, it's "open access".)

[129] Two people did drown there when the tsunami initially flooded the site.

tanks boiled away[130] after which that fuel's zircalloy cladding became hot enough to melt, react with steam and generate hydrogen gas which accumulated at the tops of several tightly sealed buildings and subsequently exploded. Those chemical – not nuclear – explosions didn't actually injure anyone but them along with the fact that some of the fuel itself melted down prompted Japan's authorities to grossly over-react[131] thereby causing panic, confusion, ~1600 additional deaths, and even more property losses.

As is also the case in most Western countries, any sort of "nuclear" incident causes Japan's authorities to behave like decapitated chickens. Because of this, many Japanese people have completely lost faith in TEPCO, their government, and the clean power source responsible for much of their nation's economic success. Fukushima's people are suffering not from the earthquake and tsunami, but also from extreme anxiety ("radiophobia") which generates stress and suppresses their immune systems.

The reason that that power plant's fate constitutes a geniune disaster is that Japan's economy will continue to suffer from economic fallout estimated to eventually reach one trillion USD. That money will go to pay for cleanup, the additional high-priced imported fossil fuels that they will continue to have to pay for, loss of economic productivity, insurance claims, lawsuits and addressing stress-related health issues (including suicides) engendered by forced dislocation and fear mongering.

Japan will not meet its greenhouse gas emission reduction targets and will become more polluted because it'll burn more coal and other fossil fuels to generate electricity. Because coal concentrates natural radionuclides, it's

---

130 The removal of fission product decay heat is a key feature of any reactor's primary cooling system. When fission stops for any reason (the reactor is no longer "critical"), fission product decay continues to generate a substantial amount of heat energy. At the moment of shutdown, it amounts to about 6.5% of the reactor's full power rating, but drops to about 1.5% within an hour. After a day, it's fallen to 0.4%, and after a week it's only about 0.2%. Nevertheless, cumulatively that's enough heat to melt the fuel assemblies unless it is safely dissipated. Fukushima's subsequently melted-down reactors were generating about 1.5% of their rated heat output when the tsunami disabled their cooling systems. Had there been several Olympic-sized swimming pools worth of fresh water stored upon the hill behind them they would have survived intact.

131 "Over reacted" means that many of the official decrees negatively impacting the Japanese peoples' lives and livelihoods were absolutely unnecessary – see this chapter's next sections. The only thing that really needed to happen rad-wise was for those authorities to assure that the locals consumed a total of roughly one gram (less for infants) of sodium or potassium iodide during the first month after the tsunami [185]. Some of their actions suggest that they believed that radioactive atoms are like pathogenic microorganisms; i.e., a trace of contamination might multiply within its victim rendering him/her a potential "carrier" of something deadly that other people might catch.

likely that the average Japanese person will be exposed to slightly more radioactivity than when all of their nuclear plants were running.

Japan will move to renewables more quickly than planned and, because its feed in tariff for them is high, that will also raise its citizen's electricity costs.

World wide, it's been an even bigger disaster for the same reasons.

The reasons for that power plant's dismal fate are the same human factors responsible for most of the other "issues" prompting today's widespread distrust of nuclear power. In that particular case, human nature first manifested itself in reactor siting. GOOGLE EARTH reveals that the Fukushima Daiichi industrial complex occupies a roughly one square mile parcel of coastland ranging in altitude from 0 to about 40 meters above sea level. Its reactors and their attendant spent fuel storage buildings occupy roughly 10% of that area and are all situated along the coast line where the ground is currently only about 10 meters above sea level – ancillary buildings etc., are situated on the shallow "hill" behind them. Since the tsunami that did the damage was "only" about 13 meters high, one of the ways that that disaster could have been averted would have been to simply site those reactors upon that parcel's higher areas rather than immediately next to the ocean. Another would have been to build a higher sea wall.

Documents filed with Japanese authorities in 1967, show that when TEPCO (the electrical utility owning the facility) was planning its new nuclear power plant, it decided to deliberately lower that site's natural, 35-meter high seawall (cliff) to just ten meters which, of course rendered the power plant built upon it vulnerable to the 13 meter tsunami that struck it in March 2011. A former TEPCO executive who had been part of that decision making team explained that it was based upon two considerations. One, that reducing that site's natural cliff wall's height by 25 meters would render delivering equipment to it much easier (equipment which was mostly delivered by sea – a full sized nuclear reactor's Rankine-type turbo generators weigh about 500 tonnes); and, two, that it would be somewhat easier/cheaper to access the seawater which would be cooling its reactors from 10 meters above sea level rather than 35. In 1991 the USA's Nuclear Regulatory Commission warned Japan's decision makers of a risk to that site's losing its emergency power system (its big/heavy backup diesel-powered pumps were situated in basements where they could be readily flooded) and Japan's own "Nuclear and Industrial Safety Agency" (NISA) subsequently reminded them of that again in 2004. Also in 2004, two more of Japan's governmental oversight committees issued warnings that tsunamis over twice the height assumed by TEPCO's planners circa 1967 (5.7 meters) were possible. In response to the Indian Ocean tsunami, during

2006 and 2008 two TEPCO-employee studies investigated the effects of tsunami-waves higher than their facility's "official" (design basis) wave height upon its performance. The 2006 simulation's conclusions were that a 13.5 meter wave would cause a complete loss of all power and render it impossible to inject water into reactor No. 5 and that the costs of protecting the plant from them would be about 25 million dollars (under 0.01% of what Japan's ministry of trade, and industry ministry subsequently declared (in 2016) that dealing with the subsequent disaster would end up costing). The 2008 study assumed a more moderate 10 meter high tsunami. In both cases TEPCO failed to act because its decision makers considered those studies to be training exercises for junior technical employees and neither they nor their apparently "captured" regulators really expected such large tsunamis.

The similarly designed and sited Fukushima Daini was also successfully shut down but suffered virtually no damage.

That fiasco generated plenty of opportunities for fear mongers to scare people about Fukushima's "deadly radiation". The goal sought and often attained by such mongering is to scare innumerate (normal) people with big technical-sounding numbers and acronyms. For example, a recent Dailykos posting's really big scary number is that "*some of the tuna caught near Japan had up to 1000 Bq*[132]*/kg of deadly* ^{137}*Cs in them*".

Let's look at that FUD quantitatively assuming a victim both rich and dumb enough to eat nothing but Japanese tuna; i.e., those fish most exposed to the Fukushima reactors' "terribly radioactive" drainage. Let's also assume that that person weighs 70 kg and needs about 2000 kCal per day to keep body & soul together.

Since 3.5 oz of tuna has about 200 kCal, he'd/she'd have to eat about 362 kilograms of it every year.

That's a tuna consumption rate of 1.15E-5 kg/sec which at 1000 Bq/kg equates to 0.0115 Bq of ^{137}Cs consumed per second.

According to the radiation decay equation (see APPENDIX XI), that corresponds to consuming 1.58E+7 [0.0115*3600*24*3600/ln2] ^{137}Cs atoms/sec.

Over a year, that'd be an accumulation of 4.98E+14 ^{137}Cs atoms if 100% of 'em "stuck". However they wouldn't stick because Cs has a biological

[132] Bq (one of the smaller measures of radioactivity in common use) = 1 decay "event" per second = 1 d/s = 1 dps. One "Curie", another common unit, = 3.7E+10 Bq. An expert that really wants to scare you, might express radioactivity in terms of pico or fempto curies (the potassium within an average-sized banana subjects anyone eating it to ~600,000 fempto curies worth of beta-type radiation for several weeks!! A can of beer is only about one fourth that deadly.)

half-life of ~70 days (less if one drinks lots of beer) which means that ~0.27% (0.5^(365/70) of the ^{137}Cs atoms consumed would actually contribute to his/her rad dose– that's an average of 1.34E+13 "evil" atoms decaying away within our hypothetical maximally exposed victim.

The radiation decay equation equates that many ^{137}Cs atoms to a decay rate of 97,900 disintegrations per second (Bq) throughout that year.

Since each ^{137}Cs decay generates about 1.1 Mev & 1 ev = 1.6E-19 J, that's a whole body dose of 0.0544 J/year which in a 70 kg body comes to a dose of 0.0077 J/kg (Sievert) [97000*1.1E+6*1.6E-19/70].

The average US Citizen gets about 0.0062 Sievert's worth of radiation per year – more if he/she lives in an especially dangerous place like western Colorado or spends much time jet-setting or mountain climbing.

Consequently, our extremely rich (that much tuna would cost ~10x an average Japanese citizen's disposable income) & maximally-exposed victim would be getting about the same dose from it that average US citizens normally receive from the part of world that they live in.

7.3 The Nuclear Industry's LNT-Based Radiation Dose Assumptions

The "technical" rationale underlying both the USA's radwaste boondoggling and Japan's response to TEPCO's travails is that the international body responsible for setting radiation dose recommendations (ICRP) utilizes an overly simplistic linear dose effect model assumption to predict the effects of vanishing small radiation exposures upon living creatures. That model's acronym is LNT which means linear-no-threshold. Its principle was introduced in the late 1950s and remains the basis for dose limits recommended everywhere because it's both politically correct and conservative. Back in the 1950s a "target theory" dominated the evaluation of radiation data. It was assumed that the biological impact/result was due to a "hit" – and the number of hits does indeed increase linearly with dose. A few years later it was also accepted that the initial step for carcinogenesis (cancer's initiation) is damage to a cell's DNA. Again, the initial number of DNA-damages does increase linearly with dose. Thusly as long as our experts and authorities choose to ignore the defense and repair mechanisms possessed by any/all of the Earth's real-world biological systems, it's not unreasonable (conservative) to assume the LNT-model. It's especially useful to anyone seeking to frighten people or rationalize additional personnel expenditures (more "health physicists") and/or super-profitable radwaste boondoggling.

The fact is that there are dose thresholds below which the cell repair mechanisms protecting us from the free radical damage caused by the fact that we eat food and breathe a reactive gas (oxygen) also protect us from the free radicals generated by low dose rate ionizing radiation (certainly anything under about ten times above normal background). That's why the nuclear industry's workers live at least as long (actually longer) than do those of most other professions, and there have been no massive die offs after the initial effects of Hiroshima's bombing, the USA's atmospheric bomb testing, or even big screw-ups like Fukushima's have dissipated.

For example, in the southwest Indian state of Kerala, despite background radiation levels about 30 times the global average, children under five have the lowest mortality rate in that country and an average life expectancy of 74 [186].

For thousands of years, Keralites have lived with background radiation levels three times that which triggered the evacuation at Fukushima, where the limit was (on July, 2016) 20 mSv/yr - the same figure permitted for most of the world's professional radiation workers. In contrast, sections of Kerala experience an average of 70 mSv/yr, with some areas as high as 500 mSv/yr and their locally grown food averages about five times as radioactive as is that consumed in the United States. Despite its high background levels, Kerala's cancer incidence is the same as that of greater India, about one-half that of Japan's and under a third that of Australia's. As the linked article says, "*Cancer experts know a great deal about the drivers of these huge differences, and radiation isn't on the list.*"

A similar situation obtains in Ramsar, a city in northern Iran, whose citizens have received an annual background radiation dose of up to 260 mSv/y for many generations – that's 13 times higher than that which triggered the expropriation of ~100,000 of the Fukushima Prefecture's citizens' homes, farms, and businesses by their government. Here's part of the ABSTRACT of a paper describing how Ramsar's elevated background radiation impacts its people's health [187]

> *"Cytogenetic studies show no significant differences between people in the high background compared to people in normal background areas. An in vitro challenge dose of 1.5 Gy of gamma rays was administered to the lymphocytes, which showed significantly reduced frequency for chromosome aberrations of people living in high background compared to those in normal background areas in and near Ramsar. Specifically, inhabitants of high background radiation areas had about 56% the average number of induced chromosomal abnormalities of normal background radiation area inhabitants following this exposure. This suggests that adaptive response might be induced by chronic exposure to natural*

background radiation as opposed to acute exposure to higher (tens of mGy) levels of radiation in the laboratory. There were no differences in laboratory tests of the immune systems, and no noted differences in hematological alterations between these two groups of people."

Note that this study's authors observed the same effect that other such studies noted; i.e., that modestly elevated radiation levels seem to strengthen human immune systems akin to the way that physical exercise does.

The USA's hyperconservative approach to establishing radiation guide lines impacts many of its citizens too. For instance, an abandoned elemental phosphorus plant situated ten miles west of Pocatello ID is shadowed by a roughly 30 million tonne mountain of phosphate rock slag that the EPA has deemed too radioactive to utilize for any constructive purpose. Prior to that decision it had been widely used for road and parking lot construction, construction fill, railroad ballast, home foundations, driveways, and even builtup-type roofing materials. The "alarm level" imposed upon the people owning, living in, or trying to sell such property (20 mrem/hr or 0.00175 Sievert/a) represents just ~25% of the radiation dose that an average American citizen gets from everything else he/she does [188].

There are a number of other well-written papers and books freely available to anyone wishing to learn more about this subject [189–192].

The elements responsible for most of our background radiation (uranium, thorium, and potassium) render "advanced" life possible on Earth because they generate the heat necessary for two of its unique features: plate tectonics and a still molten and therefore magnetic iron/nickel core [193].

7.4 ALARA (As Low As Reasonably Achievable)

ALARA is another fine sounding principle that's served to cripple the nuclear industry, especially anyone working for/at the USA's national laboratories[133]. In practice, the "reasonably" part of that acronym is ignored which means that it really says that there are no limits below which additional expenditures/efforts/training to further reduce personnel radiation exposures aren't always "worth it". That principle tacitly assumes both

[133] Since DOE's laboratories don't have to actually produce anything other than "reports" there's no real-world pressure upon them to do anything quickly and efficiently. To the contrary, foot dragging keeps their missions alive which, in turn, keeps the study-money coming in.

that the LNT hypothesis is correct and that the nuclear industry and its helpers will always have cost plus contracts and hence never face serious competition regardless of how much additional overhead costs they choose to assume. Neither assumption is correct. Since the "safest" way to work with anything that's radioactive, is to not work with anything that's radioactive, the DOE nuclear complex's leaders have replaced most of its "hot" laboratories/pilot plants along with the people that had worked in them with computerized modeling/modelers thereby rendering their laboratories incapable of doing the real world experimentation required to develop anything genuinely new.

> "*The country needs and unless I mistake its temper, the country demands bold persistent experimentation. It is common sense to take a method and try it. If it fails admit it frankly and try another. But above all try something*". Franklin D. Roosevelt.

The main reason that many people fear radiation is a deeply rooted psychological fear of mysterious entities that can't be seen or touched (e.g., Blair-type witches and voodoo)[134]. Such fears were originally engineered into our ancestors' more primitive brains on the African savannah millions of years ago to make them safer from predators at night. However, we're not quite so primitive now and should know better than that: we've all seen images of the mangled and burned bodies caused by car accidents, yet we continue to embrace automobiles because such risk is considered worth taking because we have trained both ourselves and our children to distinguish between rational expectations and gut feelings. We must to become willing to do the same with nuclear power, especially with new reactors featuring improved safety, lower costs, and greater efficiencies.[135]

---

[134] During the last millennium, roughly a million certified "witches" were executed because they could not prove that they had not caused harm to someone or something. In the same way, since one cannot prove that tiny amounts of radiation did not cause a particular leukemia—for that matter one cannot prove that they caused it either—those who wish to succumb to low-level phobia do so, those who don't wish to do so, don't. On the other hand, very few people fear the smoke emitted by burning some fossil fuels and most sorts of biomass currently responsible for about 7 million deaths per year. Consequently, James Hansen has estimated that nuclear power has thereby already saved about 2 million human lives [194, 195].

[135] "Nothing in life is to be feared, it is only to be understood. Now is the time to understand more so that we can fear less" Marie Curie.

7.5 Over Blown Proliferation Concerns

Many of the West's nuclear decision makers refuse to consider anything that might lead to greater "proliferation risk". For instance, they might say that, "*your reactor concept/scenario is impossible because its fissile isn't 'denatured'*[136] *enough.*"

That assertion makes about as much sense as claiming that it would be impossible to revive the USA's space exploration program because its decision makers would prefer that NASA's technical initiatives be powered with conservation. First, since it tacitly assumes that any new reactors would be subject to the same set of arbitrary (and apparently immutable) man-made rules as are today's, it also tacitly assumes that the world will always need the uranium enrichment facilities that represent a far greater proliferation threat than does the fissile material within any nuclear reactor. Second, diluting ("denaturing") the ^{233}U in, let's say, the MSFR or LFTR's salt streams with ~8x as much ^{238}U would: 1) render its fuel cycle unsustainable and therefore obviate the main reason for implementing any sort of nuclear renaissance; 2) greatly complicate that reactor's operation thereby increasing its electricity's cost (and, likely also, its operators' radiation exposures); and 3) turn that resulting converter-type reactor into just another large-scale transuranic radwaste (plutonium etc.) maker. It's also hypocritical because the US federal government – a signator and vociferous proponent of the Nuclear Non-Proliferation Treaty – has been operating many HEU-fueled (naval) reactors for over a half century and is likely to continue to do so. There have been no "diversions" of their bomb grade fissile by terrorists and it's unrealistic to assume that any within a power reactor sited within any first world country would be either.

Any uranium fueled reactor breeds at least some plutonium[137] which could in principle be recovered and perhaps – if it's >90% ^{239}Pu, not reactor grade plutonium – could become the "pit" of a nuclear weapon. Some folks therefore claim that any sort of fuel "reprocessing" (an absolute necessity for any sort of genuinely sustainable nuclear fuel cycle) would dramatically increase the risk of nuclear war. It's true that it's "possible" but is also very

---

[136] Denatured means that the actinide isotope(s) constituting its fissile (e.g. ^{235}U) is diluted with enough of that element's non fissionable isotope (s) (e.g., ^{238}U) that it would be impossible for imaginary terrorists to make a bomb of "diverted" reactor fuel uness they were also able to "enrich" it. The degree of dilution required is <20% for ^{235}U and <12% for ^{233}U.

[137] About 40% of a typical uranium-fueled LWR's heat energy is generated by *in-situ* generated/burned ^{239}Pu.

much less likely than, let's say, runaway global warming due to our refusal to stop such dithering or North Korea's (which has no power reactors) leadership's deciding to nuke US military bases in South Korea and Japan. Choosing to develop a nuclear arsenal is a political decision, not a technological imperative. All of today's technologically advanced nations are capable of developing/building nuclear weapons if they choose to do so. There are about 40 such nations in the world today and only nine of them have built/maintain nuclear arsenals. The others chose not to do so and most have signed a solemn treaty to that effect. Nations and people don't have to act stupidly and/or selfishly – they choose to do so[138].

[138] For instance, North Korea, a poor nation refused admission into the "nuclear energy club", built its own tiny (5 MW) "production" reactor to make plutonium bomb pits with. The much richer South Korea generated several thousand times as much nuclear energy to power its economy but refused to make bombs.

8

"The Damned Human Race" [196]

The biggest barriers to implementing my scenario's technological fixes are posed by human not Mother nature. Unlike the "controlled fusion" will-o-the-wisp that has dominated DOE's nuclear reactor development efforts/funding for several decades, there is almost no doubt that a sustainable molten salt fission-type reactor could actually be built. The USA simply hasn't been willing to do the real world experimentation required to work out that scenario's implementation details.

Solving the tough technical problems posed by the natural world's rules requires intellectual honesty – seeking Nature's opinion regardless of whether or not it agrees with yours, your peer group's or your boss's personal beliefs. It's about finding reasonable solutions, not winning arguments or pleasing the people that fund you. It's also sometimes hard and even somewhat "risky" work. To many of the people who've chosen to become scientists & engineers, doing such work also happens to be a great deal of fun[139].

For success to happen, R&D workplace managers must foster a culture where everyone can:

- Feel safe to speak up
- Express alternative points of view
- Challenge the status-quo and/or their bosses
- Acknowledge mistakes without fearing punishment

Unfortunately, that would require those leaders to embrace vulnerability, which is much easier said than done. It's tough for the most "important" people within any organization to let go of their need to be, and, equally

[139] With a few exceptions I don't miss the people that I worked with at INL but do miss my lab (it's been "decommissioned" along with almost all of INL's other nominally hot labs). To people like me, a well stocked laboratory is a wonderful playground. "*Happiness lies in the joy of achievement and the thrill of creative effort.*" Franklin D. Roosevelt.

Darryl Siemer (ed.) Nuclear Power: Policies, Practices, and the Future, (147–164)

important, to always have been[140], "right" about everything. Decisions should be grounded upon facts, not by schedules or the stature, position, or history of individuals working on the problem.

System justification theory (SJT) explains why people within large groups often act/vote in ways that are directly contrary to their own self-interests. It points out that system-justifying beliefs serve a psychologically palliative function to many people even when the system in question is disadvantageous to them. It's responsible for the counter-intuitive behaviour often seen in oppressed ("low-status") groups, wherein they buy into their oppressors' propaganda.

According to SJT, humans have three competing psychological motives: ego motive, group motive, and system justification. Our ego says, "*I like me.*" The group motive makes us say, "*I like us*" (and, conversely, "*I don't like them*"). However, the system justification motive causes us to believe that, "*I like things the way they are.*" Sometimes these motivations reinforce and sometimes they compete, depending upon circumstances.

System justification embodies our desire to believe that the world is logical (makes good sense) - that things are the way that they are for a damn good reason (God? the Constitution?). It is why many low-status people within a culture often defend the privileges and sometimes grossly inappropriate behavior of its high-status people. Further, it explains why if disadvantaged members of a group are not having their ego or group justification needs met, they will actually be less likely to demand societal/cultural changes. They rationalize their position within it, and internalize a belief that they deserve to be in their current position (*I'm just a chemist, who am I to question what the Chem Plant's managers want to do?*). Of course, the advantaged members of such groups have an even more "conservative" view of the world: "*things are right as they are*". Consequently, to them, everything is simple: just set up the rules and structures ("procedures") to be blind to individual characteristics, and the system will assure that everything will work out fine & everyone will have an equal shot at happiness and success. In such a culture, if anyone happens to fail, it's due to a flaw in their personalities, not the system itself[141]. In many ways,

---

[140] This is one of the reasons why institutions rarely start over from scratch when projects like this book's examples of boondoggling have repeatedly failed – doing so is a tacit admission that that their leadership's approach to decision making is flawed. Lower ranking people just keep pretending that their "emperors new clothes" continue to be as beautiful as they were at first sight.

[141] That's why it's so easy to brush off the complaints of such losers by simply calling them "disgruntled employees". I suspect that some of the Jews sent to Auschwitz were also disgruntled.

conservatives see "systems" to be the most important thing which is why the leaders of political parties, corporations, and institutions behave the way they do.

The examples I've already described reflect several of human nature's less admirable characteristics including...

8.1 Greed

Doing the right things at Fukushima would have cost a bit more more and thereby reduce profit margins. Doing the wrong things with Hanford, INL, and SRLs' reprocessing waste increased profit margins by making more "work". In cost-plus government contracting, no mole hill is too small to become a mountain if the people authorized to climb it choose to pretend that it is[142].

Organized (usually corporate sponsored) greed seems to have become especially pernicious. Acquisitiveness is the chief driver for entrepreneurial activity and therefore should be, but often isn't, trammeled by rules/regulations that serve the common good. For instance, in the 1970s, continued nuclear power development was still seen by most of the first world's decision makers as essential to serving the energy needs of an increasingly affluent and quickly growing population. However, uranium supplies were suddenly threatened by the imposition of a protectionist U.S. embargo that compelled its power utilities to use only U.S. sourced uranium. That embargo was imposed in spite of the fact that some of those foreign mines had been developed with U.S. government encouragement. It depressed prices from other sources, threatening to force closure of some of Canadian, Australian, and South African uranium mines. In response, the Canadian government led in the creation of a world uranium cartel supposedly so secret that any Canadian who talked about it could be jailed. Its existence was first brought to public attention when confidential documents were stolen from the files of an Australian company in 1976. It's another sad story too long to detail here (see [197]) but the point I'm getting at is that

[142] One of the things that any agency/company/government wishing to mountain-build must do is to choose "regulators" that won't interfere with its plans. In the DOE Complex, one such group is the INL's DOE-funded, Site Specific Citizens Advisory Board. It's not really their fault though because DOE can block anyone it chooses from membership on that board and everyone so considered must promise to go along with whatever its leadership desires – all of its advice/conclusions must be "unanimous". Furthermore, DOE doesn't have to act upon its advice.

that OPEC-like cartel quickly succeeded in pushing the price of uranium up by ~700% thereby destroying any confidence that the world's nuclear power utility owners may have had about the future price of their fuel.

Cornering the market on anything from filling teeth to supplying government-mandated "EpiPens" remains as much of a goal of free enterprisers now as it did then. Implementing a "Clean New Deal" represents a potentially much more important/lucrative business opportunity than is the making of EpiPens or insulin and thereby provides even more temptation for selfish behavior. This is another reason why a nuclear renaissance should not require a fuel that could be so controlled again. Natural uranium and thorium are so abundant that if they could be efficiently burned, no one could corner the fuel market.

Finally civilian nuclear power has been subjected to "technological lock-in" for many decades because its first movers succeeded in establishing a profitable business model that didn't emphasize efficiency or long term sustainability. It worked for quite a long time but its weaknesses eventually rendered nuclear power much less attractive than it should/could be (see [89]).

8.2 Tribalism

That's "paying attention only to those who think as we do and considering anyone else an "outsider" to be ignored, a.k.a., 'groupthink'": Examples include management vs workers, rich vs poor, businessmen vs technical nerds, experts vs the "dumb public", regulators vs businesspersons, INEL's vs Hanford's radwaste experts, the Chem Plant's LDS vs non-believing employees[143], Democrats vs Republicans, Shia vs Sunni, almost everybody vs whistleblowers, and the majority of "environmentalists" vs anyone that's pro nuclear. Why does Africa now have 54 different countries? For that matter, why does the USA have 50 no-longer-very-united states?

The western world's currently rising tide of tribalistic nationalism brings out the worst in us. Humans like chimpanzees are apparently hardwired to categorize others of our own species as "thems" not "us" triggering the hate, fear, disgust, violence, selfishness, etc. that we're seeing more of now than since the end of WWII [198]. Solving the world's biggest technical problem (replacing fossil fuels) will require its people to cooperate, not retreat further behind their own borders under the goading of populistic

[143] "*When fascism comes to America, it will be wrapped in the flag and carrying a cross*", Sinclair Lewis.

leaders taking advantage of our all too human nature. It can be overcome by the right sort of leadership but rarely is.

Diversity of thinking is vital if a group facing a complex problem wants to arrive at a competent solution.

Disciplinary (tribal) boundaries often hinder scientific progress because each of its discipline's adherents resent attempts by Jack-of-several-trade outsiders to either poach upon their bailiwick or evaluate their efforts. This often manifests itself as difficulty in getting anything that might be considered "controversial" accepted/published in another specialty's peer-reviewed journals[144]. Another very real fact is that scientific journalism's anonymous reviewers are simultaneously free to turn down such papers (or research proposals), and then subsequently "discover" the concept and then study/describe it themselves[145].

Most people are reluctant to eschew the beliefs of their peer group, regardless of how unsubstantiated they may be. For example, during the early 1980's Argonne got a head start on what came to be considered its Generation IV project at DOE's Idaho site. Within about five years they'd become ready to demonstrate how safe their liquid metal cooled fast breeder reactor would be. Scientists and engineers from around the world gathered to watch what happened when Argonne's engineers deliberately caused a total loss of coolant (liquid sodium) flow to EBR II. Exactly as

[144] I'll probably go to my grave with <100 peer-reviewed publications because I'm no longer convinced that publishing in professional technical journals is worth doing for people in my position (both retired & unconnected). You must write in a stilted fashion carefully avoiding anything that's too controversial, there are now too many journals, and, finally, the "best" ones typically charge their contributors several thousand dollars/paper and then turn around and charge anyone who wants to actually read what you've written, $35-$50/article. Since I now have to pay my own bills, that business model means that I rarely read anything in toto that isn't published "open access". Thanks to the internet (GOOGLE, Engineering toolbox, WIKIPEDIA, RESEARCHGATE, Science Daily News, etc. plus some still "connected" friends) and the fact that some papers are indeed published "open access", it's now possible for people like me to pursue science as a hobby. It's been fun.

[145] That happened to me while I was an "assistant" tenure-track professor of chemistry at Marquette University. During my second year there, I'd written an NSF proposal having to do with atomic fluorescence spectrometry, a field in which there was only one active research group in the US. It was turned down of course and I discovered two years later that that group had suddenly decided that it'd be a fine topic for one of its PhD candidates to work on. He just happened to be one of the people interviewed to fill my position at Marquette after I'd decided to quit (after four years) and go to work at INEL (he admitted that they'd reviewed my proposal). That's one of the reasons I quit.

predicted, its core temperature briefly increased and then rapidly dropped as it safely shut itself down without further intervention.

Although the IFR project was a technical success in some respects, it couldn't buck the USA's anti-nukes & thereby address the problem serving as its raison d'etre. The majority of environmentalists (one tribe) reflexively oppose nuclear power along with many of the Democratic Party's "liberals: (another tribe) - so the Clinton administration shut it down because "*we don't need it*" even though its leader (and certainly his vice president) were aware of the issues posed by anthropogenic climate change. The same thing happened at the beginning of the Obama administration – funding for the GEN IV nuclear reactor R&D (GNEP) program instigated by President G.W. Bush was axed[146] and the US taxpayers' thirty-year/$15 billion dollar investment in DOE's YM spent fuel storage site[147] was also wasted.

This was not surprising for several reasons, one of which was that two of the people responsible for authoring MIT's seminal. "*The Future of Nuclear Power*" (MIT 2003), Professors John Holdren and Ernie Moniz, subsequently served as President Obama's technical/scientific advisors a half decade later. Three of that report's conclusions were as follows:

- *"Placing increased emphasis on the once-through fuel cycle as best meeting the criteria of low costs and proliferation resistance;*
- *Urging DOE to establish a Nuclear System Modeling project that would collect the engineering data and perform the analysis necessary to evaluate alternative reactor concepts and fuel cycles using the criteria of cost, safety, waste, and proliferation resistance. Expensive development projects should be delayed pending the outcome of this multi-year effort.*
- *Giving countries that forego proliferation- risky enrichment and reprocessing activities a preferred position to receive nuclear fuel and waste management services from nations that operate the entire fuel cycle."*

Basically, what that report's bullets suggest is that President Obama (a lawyer by education) was told by two of this country's best connected

[146] Another reason for that cut-off was that the special interest-driven (banking) investment rule changes (regulation removal) introduced by the previous administration had led much of the world into a deep economic recession which, of course, affected governmental spending everywhere.

[147] It's likely that if everyone had simply agreed to call YM a "monitored retrievable radioactive waste storage facility" rather than a "repository", it would be serving that purpose and thereby providing several hundred good-paying jobs for Nevada's "middle class".

technical experts that developing a sustainable nuclear renaissance wasn't really necessary. Since nuclear power represented just one part of a temporary "bridge" to the future[148], "all of the above" (?) plus a few small modular conventional-type reactors *seemed* to represent the best way to address the USA's future energy issues.

In a finite world, no person or country can afford to do "all of the above"– choices must be made and a project's success/failure will depend upon whether or not they were the correct ones. Fossil fuels regardless of how they are obtained do not represent a sustainable energy source, biofuels cannot replace fossil fuels, wind/solar power is and will remain ineluctably unreliable, and batteries are and will remain too expensive for anything but niche applications. Mr. Obama's DOE didn't make much progress because its leadership refused to think quantitatively and make tough choices. Consequently, he/they got it backwards: fracked gas/oil and and today's renewable energy sources should be considered temporary bridges to Hubbert and Weinberg's nuclear-powered world.

8.3 Gullibility

"Willingness to embrace logical fallacies"; e.g., "appeal to authority", "appeal to fear" (it's no secret that in US industries, poor team players experience "solitary, poor, nasty, brutish, and short" careers[149]);" wishful thinking" (see my discussions of the consequences of USA's reprocessing waste management decisions); "sunk cost" (decisions tainted by an accumulation of emotional investment – the more already spent upon something, the harder it becomes to abandon); and finally, "appeal to ignorance"; e.g., "*extremely low level radiation dose rates must be evil because you can't prove that they aren't*" (it is statistically impossible to prove such things).

The USA's "conservative" dominated federal government's manifest incompetence has recently inspired a number of liberal young & not so young politicians to espouse a "Green New Deal" calling for a transition

---

148 The other and much bigger part of that temporary bridge would be the tight gas & oil fracked out of the USA's shale and sandstone formations.

149 One of Admiral Rickover's especially sapient quotes is, "*If you're going to sin, sin against God, not the bureaucracy; God will forgive you but the bureaucracy won't.*" Fear of job/ career loss is a pretty effective motivator for DOE's "technical" people, especially those who still have mortgages & dependents to support. So too is the fact that its trouble makers are "disfellowshipped" until their managers come up with a plausible-seeming excuse for reorganizing them out of their jobs (e.g., their expertise won't be included in the next project's personnel "skills mix" requirement).

away from a world rimarily powered with fossil fuels to one powered by Dr. Jacobson *et al.*'s 100% wind, water, & solar (WWS) scheme.

In most Western countries it's the "progressives" and socialists who claim to care more deeply about the environment that seek to shut down nuclear plants, not their mostly-conservative, climate skeptics. It's illogical because two of nuclear energy's greatest successes are Sweden and France, nations regarded by many liberal-leaning people as the best kind of societies. Through a combination of sensible policies and free-market incentives, starting circa ~1970 Sweden cut its per capita GHG emissions by a factor of three while doubling its per capita income and providing even more of its notoriously "pro-people" social benefits with nuclear power. A new Swedish book's subtitle contends that it, much as France had almost done before, has already "solved climate change" by expanding its electrical supply with nuclear power [199]. Sweden's primary concern at that time wasn't global warming but reliability: its leadership deemed further hydropower development too environmentally impactful and that decade's OPEC-driven oil crises had rendered fossil fuels too unpredictable. Consequently, between then and 1990, Sweden built a dozen nuclear power plants on four different sites, eight of which continue to operate today. They currently supply ~40 percent of Sweden's electricity which is as much as do the hydropower plants that Mother Nature liberally endowed both it and Norway with the potential to employ. Consequently, electricity in both of those countries is still cheap, clean and reliable. The authors of that book contend that, "Sweden became the most successful country in history at expanding low-carbon electricity generation and leading the way in addressing climate change". They also say that, "*without growth in nuclear power, replacing fossil fuels with renewables simply decarbonizes the existing supply. It doesn't deal with the increased demand…*".

In the real world, it's been nuclear energy, not solar and wind that has decarbonized energy supplies while increasing wages and societal wealth.

And it is only nuclear that has powered high-speed trains everywhere from France to Japan to China, thereby decarbonizing transportation – the source of about one-third of anthropogenic GHG emissions. Consequently, it seems reasonable to expect that a "Green New Deal" incorporating nuclear power is apt to be more successful and certainly less environmentally impactful than one that doesn't.

I suspect that one of nuclear renaissance's most "evil" characteristics as far as many of the green movement's true believers is concerned is that it would not require their fellow citizens to embrace their faith in non nuclear renewable power – most people could just go on living in pretty much the same way that they've gotten used and not have to worry about whether the power might go down ("load shedding") in the middle of whatever they're trying to do[150].

8.4 Laziness

"Unwillingness to work": for decision makers that means not taking the time to learn the fundamentals of what you are supposed to be managing. Ignorance – especially if it's deliberate – should never represent an excuse for any failed project's leadership but often does[151].

Because DOE's "technical" decision making is primarily driven by immediate political considerations and schedules, its top level managers rarely bother to acquire much relevant "subject matter expertise". The people who provide them with that service – mostly senior contractor employees – are rewarded for saying/writing/supporting whatever "the customer" happens to want at any given point in time; e.g., an excuse for espousing "separations" instead of »direct vitrification» of INEL's radwastes - and eventually downsized if they persist in not becoming good team players[152].

The funniest example of this that I can remember was back when we lab rats had just been informed that INL's (then INEL's) "new mission" would be to become the "lead lab" in radioactive waste management and DOE's brand

[150] Coal power even in coal-rich countries isn't necessarily very reliable [200]. South Africa's Eskom – the African continent's biggest power utility by far - has just decided to take 4,000 MW off its power grid. On 14 March 2019, the first 800 MW unit of its Kusile power plant tripped, exacerbating a shortfall of generating capacity and prompting it to announce a Stage 2 (2 GW) load shedding, that changed into a Stage 3 (3 GW) and finally into a Stage 4 (4 GW) load shedding, after the loss of additional including 900 MW imported from Mozambique due to storm damage to an interconnection to its Cahora Bassa hydropower plant). Its existing plants suffer many breakdowns and the construction capacity, of two more big ones (4.8 GW each), is running years behind schedule and billions of dollars over budget. In mid-March 2019, ~12 of Eskom's 45 GWe total capacity was unavailable due to unplanned outages and its back-up diesel supplies were also under pressure. Eskom is now US$29bn in debt and South Africa's government has announced plans to restructure it and support some of its short-term debt issues.

[151] The excuse of last resort is that a repeatedly failed project is/was "*first of a kind*". In most such cases that's true because no other institution would have initially decided to try to solve the problem that way. DOE's managers do so because they can always blame/change contractors and about the worst possible punishment that a >50 year old civil servant can suffer is a well-funded early retirement (a government contractor's personnel are not similarly protected).

[152] The most common way of downsizing troublemakers was to not include her/her expertise in the Site's next project's "personnel skills mix" requirements.

new Assistant Secretary of Energy (Management and Administration) had decided to go out and meet the stakeholders. When asked by a newspaper reporter to define "High Level Waste" everyone learned that...

> "*Well it is a radius that is applied to it on a high level in a way that you are going to treat that way, and the way that you are then going to process it. So at the end of that, on a high level, let's say you might have to – after processing that, you may-before you put it into the permanent storage, that may be something that you have to transport to let's say - to where we were putting storage in new Mexico. Some other kind of waste you could be able to put in the cement and store it right there on the site. High level, we would handle that with care. We handle all of it with care.*"

That really cleared things up for us newbies.

8.5 Deviousness

"Propensity to deliberately mislead others to serve your own purposes". In technical reports this is usually evinced by deliberately not addressing what should be the subject's "obvious" issues/questions while emphasizing the importance of whatever its author is doing/proposing. This usually "works" because the real customers for such information (funding sources, politicians, and/or important citizen stakeholders) are often technically clueless and just seeking reassurance that everything is going along just fine. The reason that DOE's nuclear arm has become dysfunctional and its documentation/advice unreliable[153] is that its people, top-to-bottom, have behaved and written the way they felt their bosses expected/wanted them to (see APPENDIX XIII). That's also why its documents and contracts undergo so many "revisions"[154].

I've told everyone who's invited me to become a member of their MSR nuclear startup team that being "sneaky" about what you are proposing to do is no longer apt to win hearts and minds (certainly not mine) – too many people have been snookered too many times by the nuclear industry's

---

153 As Sinclair Lewis said almost a century ago, "every compulsion is put upon writers to become safe, polite, obedient, and sterile".

154 For instance, after INEL's new naval fuel dissolution facility proved to be far less productive than "promised", previously published reports were revised to eliminate specific throughput estimates & the original versions gathered up & destroyed. The same thing happened with the reports having to do with its decision to "steam reform" its remaining liquid reprocessing waste.

experts. That cultural foible is what finally made me decide not to align myself with any of them[155] - protecting "IP" is just another excuse for devious behavior and likely cause of failure.

APPENDICES XIII & XIV detail specific examples of institutional deviousness.

8.6 Bullheadedness

Once we humans decide to look at things in a certain way (establish a "business model"), it's extremely difficult to change our paradigms/minds even when it becomes obvious to "outsiders" that it's not working. Here are some examples....

8.6.1 Anti Nuclear "Environmentalists"

Reflexively antinuclear environmentalists are especially bullheaded[156]. The real world limitations of wind, solar, and even hydropower[157] that most of them refuse to recognize means that every time they vote/rail against nuclear power, they inevitably vote for the fossil fuel industry, especially oil and gas fracking. It's likely that those fossil fuel suppliers are delighted since they know that for the foreseeable future, the only thing that can replace nuclear power plants shut down by such activism are new coal and natural gas-fired plants. Since "No Nukes" became a signature cause of the Green movement in the late 1970's, sympathy to nuclear power became to many environmentalists a sign of disloyalty if not treason.

Today's rabid antinukes are much like the people decades ago who were "enraged" about the environmental effects of the "green revolution" that subsequently made it possible to feed today's ~7.5 billion people. Professor

---

155 I did some work for three of them but backed out when it became apparent that their goals were inconsistent with those identified in this book.

156 Paul Ehrlich, one of the recent past's most influential misery-mongers famously opined that giving cheap, abundant energy to humanity would be like "*giving an idiot child a machine gun*". Today's fossil fuel empowered idiot children, some of whom are over 70 years old and hold high elected office, are waving around their new mini nuclear bombs.

157 One of the assumptions underlying the 100% WWS (Wind Water & Solar) scheme is that dams (that proposal's "batteries") could temporarily provide over an order of magnitude more electricity than they actually could. One problem with that assertion is that existing "good dam sites aren't big enough to accommodate an order of magnitude more power houses/turbines. Another is that such massive and sudden water releases would flood anything down steam of them.

Borlaug dismissed their concerns with, "*some of the environmental lobbyists of the Western nations are the salt of the earth, but many of them are elitists. They've never experienced the physical sensation of hunger. They do their lobbying from comfortable office suites in Washington or Brussels...If they lived just one month amid the misery of the developing world, as I have for fifty years, they'd be crying out for tractors and fertilizer and irrigation canals and be outraged that fashionable elitists back home were trying to deny them these things.*"

8.6.2 Hyper Secrecy

Even scientists and engineers tend to equate arbitrary and often poor human-made decisions with Nature's laws. Unfortunately, nuclear power was an afterthought of a post WWII rush to develop super bombs and submarines that could stay down indefinitely, move quickly, and break things. Of course, as such, everything about it/them became a secret that only experts and certain authorities could access or understand if they could see the classified reports. Secrecy within the DOE complex has been gradually getting worse ever since the end of the Clinton administration, but Mr. Trump's DOE performed the coup de grace by completely pulling its public-facing employee directory ("DOE Phonebook") behind an impenetrable firewall, thereby making it far more difficult for both journalists and members of the public to locate the email addresses and/or phone numbers of its employees both civil service and contractor. It's become almost impossible for an outsider to discover who's actually responsible for doing anything at DOE's laboratories.[158].

Rendering federal scientists and policy makers harder to contact isn't a trivial matter. Firewalls shield taxpayer-funded government employees from the outside world, meaning that they don't experience much public feedback and can therefore continue to do whatever they please. This has been getting worse since the end of the Clinton administration, which is why everything about nuclear power R&D has become even more tarred

[158] Even information obtained via a "Freedom of Information Act" (FOIA) request usually serves to protect our government's decision makers. An example of this came to light when a lawyer representing two of Idaho's ex governors (Batt and Andrus) made a freedom of information request to DOE about specific questions regarding obviously already made decisions regarding shipments of spent fuel rods to INL prior to any environmental analysis or public input: "*The response is a joke,*" Lucas said. "*They have redacted almost all their supposed answers and are claiming national security concerns on almost every question. Literally, DOE has not produced a single document that was not already public,*" *violating the Freedom of Information Act.*" [201].

with the same brush as has nuclear weapons development. A nuclear power plant can't explode like a nuclear bomb[159], but many people think that it could and, quite justifiably in my opinion, no longer trust that industry's experts and spokespersons.

The USA's determination to "classify" anything having to do with nuclear reactor development is one of the reasons why it neither leads the world in that field nor is apt to regain leadership. This compulsion has recently been rendered more destructive by its federal government's decision to pit the USA's entrepreneurs against each other to develop a renaissance that it should assume responsibility for developing[160]. Consequently, those "best and brightest" US citizens most likely capable of implementing change are forced to not cooperate with each other in order to protect their "intellectual" property. That is, of course, perfectly OK with the secrecy-obsessed DOE managers responsible for "helping" these go-getters with research grants and/or vouchers that can be spent only at their national laboratories (give their employees different numbers to charge their work time to). This paradigm assumes that people in other countries are incapable of either doing such research for themselves or even reading reports published by the citizens of countries that haven't chosen to similarly hamstring themselves.

It also happens to be absolutely contrary to the guidance that Alvin Weinberg gave his younger colleagues at ORNL fifty two years ago:

> "*Good people from diverse fields working together can make scientific discoveries that are denied to solitary geniuses working in isolation. Such coherence in applied research projects is even more important.*"

---

[159] The two reactor explosions" that I'm aware of – that of the US Army's SL-1 (Stationary Low-Power (3 MWt) Reactor Number One) and the USSR's Chernobyl disaster - were steam, not nuclear, explosions that immediately shut down both of them. For example, the deliberately caused (suicidal) "transient" that shut down SL-1 generated about 20 GW for about four milliseconds [202]. That's 8E+7 joules which means that about one milligram of its fuel was consumed during it – a small nuclear bomb's explosion consumes 5-6 orders of magnitude more fissile.

[160] The US federal government is encouraging people who grew up during a time when exploiting "intellectual property" had become the best way to become rich and famous, to decide what should be done or tried. There's a lot more to implementing a sustainable nuclear renaissance than just creating or cornering IP. Pitting entrepreneurs and/or contractors against each other to do the government's job simply sets up another situation in which the biggest liar is apt to "win". The development of a reactor/fuel cycle able to affect Weinberg's "Age of Substitutability" is too big, too important, and too potentially dangerous to turn over to people/institutions primarily motivated by personal gain/ profits or immediate political drivers. Everything we know about achieving it should immediately become "open access" and permanently kept that way.

8.6.3 My Own Bullheadedness

A final example would be my own refusal to pretend to just go along with whatever the System was apparently determined to do and then try to change things as an "insider" rather than an outsider (see APPENDICES VII & VIII). That may have even worked, who knows? For instance, I've never become a member of the American Nuclear Society despite having several papers relating to some of the technical aspects of how a fission-based nuclear renaissance might best be implemented published in its peer-reviewed journals. The main reason for this is that The ANS is dominated by people/institutions that are in one way or another, responsible for that discipline's technical stagnation. For example, the person who headed-up US DOE's nuclear engineering R&D program during the latter half of the Obama administration (it's "NE 1") was chosen to give a plenary talk at the 2017 National ANS meeting. As far as I was able to tell throughout his tenure, both he and his counterpart on DOE's "waste" side (its "EM 1"), just kept kicking their respective cans on down the same road (to nowhere) without apparent interest in learning why they shouldn't do so. People who receive such honors from organizations like the ANS tend to be exceptionally good "team players", not visionary technical experts. Unfortunately, the top-down driven institutions that employ them (DOE and its contractors and other business partners) impose blinders upon their employees which keeps them from doing anything inconsistent with that institution's immediate interests which, in turn, are driven by the policies/attitudes of the USA's political leadership.

8.7 Bullheadedness' Consequences

8.7.1 INEL's Calciner's Off Gas "Opacity Issue"

One of DOE's Idaho laboratory site's most distinguishing features was that when its nuclear fuel reprocessing plant's waste calciner was being productive, a huge, somewhat toxic[161], golden-brown plume of NOx would stretch out downwind of its ~300 ft high "smoke" stack. That plume would occasionally wander over the ~25 miles-away "Craters of the Moon" national monument & thereby engender an "opacity issue". INL's calciner operated at about 500°C and was heated by squirting kerosene along with

[161] Both its odor and effects are similar to those of chlorine gas.

the liquid waste into the bottom of an excess-air-fluidized bed reactor[162]. All of the liquid wastes so-calcined consisted of mixtures of water, nitrate salts, nitric acid, and, often, some fluoride salts too. With "high fluoride" wastes, a great deal of calcium and aluminum nitrate salts (more nitrate) were added to "complex" it (reduce corrosivity) while it was still liquid and suppress HF volatilization during its subsequent calcination. Under those conditions the majority of nitrate salts[163] and all of the nitric acid decompose to form a mix of NO and NO_2 which along with excess air, elemental nitrogen, water vapor, and CO_2 was filtered and then sent up the stack. Because a lot of the waste (anything created by reprocessing zirconium clad fuel) did indeed contain fluoride, a great deal of calcium nitrate was added most of which was imported from Sweden because it was cheaper than that from US sources. At the same time that this was going on, the site's reprocessing plant's facility's fuel dissolution experts were importing lots of nitric acid.

In the late 1980's I pointed out to my management that the NOx in the calciner's offgas would be easy to recover/purify[164] and that doing so would be advantageous to them/us in several ways. One was that because NOx is both highly visible (impossible to hide) and somewhat toxic, it was generating "stakeholder" complaints (by that time, it was illegal to emit nearly that much NO_x anywhere else in the USA). The other was that scrubbing/condensing it out of the calciner's off gas to make the nitric acid and calcium nitrate that we'd otherwise have to import would certainly be politically correct and might even save us some money. However, management didn't want to change procedures or complicate their jobs so that suggestion was ignored – the Federal government's nuclear facilities could flout its own laws then and apparently still can[165]. Similarly, a few years later when it became apparent that INL (then INEEL) should immediately cal-

---

[162] In a fluidized bed reactor, a fluid (usually a gas) is passed up through a "bed" of solid granular material (e.g., already-formed calcine) at a velocity sufficiently high to suspend those particles causing the system to behave as if it contained a well-mixed low viscosity fluid. They are used to carry out a variety of multiphase chemical reactions.

[163] Alkali metal (sodium and potassium) nitrate salts are the exception – at 500°C they melt to form a glue-like liquid that "agglomerates" (aka "rocks up") the sand-like particles comprising the calciner's fluidized bed which eventually shuts it down.

[164] Nitric acid is normally made by scrubbing an oxygen plus NO_x gas mix with water which means that the only "new" thing required would have been a more-or-less off-the-shelf gas scrubber. The INL's reprocessing facility already had a large distillation system which could have purified/concentrated such acid.

[165] Forcing people to work without pay is something that only our federal government's topmost businesspersons can do.

cine its million gallon inventory of still-liquid "sodium bearing" reprocessing waste, its management chose to shut that calciner down rather than change their procedures[166] which, eventually led to the billion-dollar SBW "steam reforming" boondoggle that I've already described.

8.7.2 Argonne Idaho's IFR Waste Management Scheme.

Nearly four decades ago the United States decided to focus its sustainable reactor development effort exclusively upon Argonne National Laboratory's plutonium-breeding, liquid metal–cooled, fast reactor. Consequently, that concept currently represents the only genuinely sustainable nuclear fuel cycle that could be implemented "quickly." Unfortunately, as developed, it possesses a number of drawbacks that have rendered sustainable nuclear power a tough sell to both electrical utility CEOs and outside reviewers.

One of those drawbacks is that the fission product containing salt waste generated by Argonne's "electrorefiner" was to be converted to a Ceramic Waste Form (CWF) rather than glass[167]. CWF is a mixture of a low melting powdered glass "glue" and a synthetic sodalite powder made from the electrorefiner's "hot" (radiologically contaminated) LiCl/KCl electrolyte plus zeolite which is hot-sintered together to form a ceramic monolith. The problem with that is that CWF is simultaneously very difficult/expensive to make, not very durable (leach resistant), and can contain only a small percentage of the fission product waste that the glass-type waste forms produced by the world's still-functioning reprocessing facilities can.

In other words, an IFR-based nuclear fuel cycle applying CWF to its reprocessing/recycling waste streams would generate 15 to 19 times more high level waste (HLW) form material/GWe-yr than do today's once

---

[166] INL's calciner was fed from a 10,000 gallon tank containing a mixture of liquid waste plus whatever else deemed necessary (e.g., calcium nitrate for fluoride containing wastes). Mixing such large volumes of ~8 molar nitrate salt/nitric acid mixture with enough sugar to prevent agglomeration could become "dangerous" if, for some reason, that mixture wasn't quickly calcined (it's thermodynamically unstable and therefore might "explode"). Of course, that's certainly not the only or logical way to add that sugar (e.g., add molasses via a "mixing tee" added to the pipe running between that tank and the calciner) but any proposal's trumped up "safety issues" provided the Site's management with a fine-sounding excuse to not change anything.

[167] Glass making (vitrification) has come out on top of every independent ("outside") US evaluation of HLW treatment alternatives save one. That exception concluded that concretes made with properly "pretreated" (?) wastes might be equally satisfactory and somewhat cheaper. The reasons for that near-consensus are that glasses are compact, sufficiently water-leach resistant, and the process that makes them (vitrification) is relatively safe and simple to perform.

through LWRs coupled to a modern Purex-based reprocessing facility. By volume, that disparity would be even greater because CWF is considerably less dense, ~2 g/cm^3, than are glasses, ~2.6 to 3.0 g/cm^3.

Turning around a situation like that requires a critical examination of all assumptions inherent to its paradigm (for CWF) in order to identify/challenge/change those rendering it unnecessarily inefficient.

In that case, the key wrong assumption was that an electrorefiner salt's waste form must immobilize chloride. That assumption is unnecessary because chloride is neither toxic nor radioactive nor difficult to separate. It is also a crippling assumption because none of the durable natural minerals that waste form material scientists set out to emulate can accommodate more than a few weight percent chloride. Since sodalite [$Na_8(AlSiO_4)_6Cl_2$] represents the best of a poor lot in that respect (7.2 wt% Cl), ANL's materials scientists set out to produce an artificial sodalite and therefore ended up with CWF [203].

My first retirement hobby project was to develop a better way to deal with such waste. That effort involved performing a basement-laboratory demonstration of how it could be treated in the same way that the USSR had adopted/implemented for that generated by its Mayak reprocessing facility: i.e., mix the waste salt with aluminum and/or ferric oxide and phosphoric acid and then vitrify it [175]. That process is simple to do, generates an exceptionally durable (leach resistant) and compact glass waste form material, and also renders the separating/capturing/recycling of chlorine easy/simple. (Recycling is the best way to deal with any sort of industrial waste.[168]) It was tough to get that paper published because its anonymous peer-reviewers were all experts involved with the development of CWF – however, that Journal's (Nuclear Technology's) editor finally overruled them and decided to accept/publish it anyway [108]. He even waived that journal's usual page charges[169].

Not surprisingly, whenever any of INL's erstwhile fast reactor development experts (most of them have either retired or gotten academic jobs) have subsequently been called upon to render an opinion (or write a paper) about how such waste is to be treated, they don't seem to remember that there's been a reasonable alternative to CWF already described/published.

---

168 For example, TERRAPOWER and ELYSIUM's MCFRs would surely run somewhat better with fuel salts containing more or less pure ^{37}Cl – not natural chlorine. Since making it will not be cheap, it should be recycled, not converted to some sort of waste form.

169 In science journalism, authors are generally expected to pay publishing houses to have their work reviewed and, if deemed acceptable, published. Both reviewers and (usually) editors are expected to volunteer their services - not paid.

The last such paper I've seen [204] recommended direct geological disposal of that water soluble, intensely radioactive [170]waste salt rather than trying to turn it into CWF (the only alternative to direct dumping mentioned was a exotic tellurite glass with about the same waste loading limitations as CWF).

8.7.3 DOE's Radwaste Classification System

Because humans invariably take far too long to change official opinions (rules and laws) when they prove inconsistent with natural law, we routinely waste lots of time and (usually) other peoples' money when things don't work out as planned. An especially silly example is the USA's radwaste classification system. Such waste comes in four flavors: commercial spent nuclear fuel (SNF), high-level nuclear waste (HLW) and transuranic waste (TRU) — both from weapons making — and low-level radioactive waste (LLW) from things like the mining, medical, and energy industries. A minor amount of other sorts of radioactive wastes are sprinkled among these categories.

The HLW currently serving to rationalize most of the DOE Complex's radwaste boondoggling is "high" because it contains some stuff originally within a waste stream (raffinate) coming out of a reprocessing facility's "first cycle" liquid-liquid extraction system (see APPENDIX I). In other words, the "highness" of a storage tank's radwaste is determined by the label assigned to the worst fraction of whatever happened to have ended up within it, not by how "hot" (radioactive) or chemically toxic it actually is. The good thing about radioactive-type wastes is that they inevitably become less radioactive as time goes by which is why much of DOE's 30 to 70+ year old production/test reactor-generated HLW is no longer very radioactive (see APPENDIX 1X). That fact doesn't matter to most of the folks involved with either performing or overseeing DOE's waste management projects – *"it's 'high' because that's what they say it is which means we're gotta (and can) treat it as if were really dangerous"*. James Conca has recently written a fine explanation of why the USA's classification system needs a radical overhaul [205].

---

[170] Such salt would be at least as "hot" as the HLW being vitrified in modern European fuel reprocessing facilities (see APPENDIX IX).

9
Why the Western World's Erstwhile Leader in Nuclear Energy Must "Embrace Change"[171]

Because the USA was virtually unscathed by WWII and had implemented several wise post-war policies, up until about 1970, it had a virtual monopoly upon both technological innvovation and industrial productivity. However, by the mid-1970s much of that war's physical devastation and attendant social and economic turmoil elsewhere had been overcome. That shifted a significant portion of the world's industrial production to Asia where cheaper labor, shipping, communication, and overhead costs gradually shifted the USA's economy from one based upon production of its own goods to one based upon services (which includes selling foreign-made goods to US customers) and consumption.

During that decade millions of foreigners willing to work for wages lower than those that its current citizens expected immigrated to the USA. In addition, millions of recently emancipated women and grown-up baby boomers entered the same labor force. Those demographic changes combined with the globalization of manufacturing and the automation of production everywhere effectively froze real (properly inflation adjusted) wages for the first time in US history. Because it had become convenient for its top businessmen to do so, the USA also voluntarily gave up its capability to provide some of its people's needs with locally-sourced mining, manufacturing and agriculture—or even to extend the life of such products through reuse, repair, and repurposing. For example, the rare earth elements (REE) needed to make things like TV screens, computer hard drives, hybrid car motors/batteries, wind tower generators, etc., are

[171] This section begins with a fairly extensive rewrite of "America's beleaguered middle class", an article originally written by Richard Larsen & published by the "Idaho State Journal", February 22, 2014. However, the solution to their/our problems that Mr. Larsen had proposed is different than mine.

Darryl Siemer (ed.) Nuclear Power: Policies, Practices, and the Future, (165–178)

no longer produced in the USA because the ore bodies containing them also contain enough of the chemically-similar thorium and its "daughters" to render them slightly radioactive, which rendered working with them in its regulatory environment prohibitively expensive[172]. We no longer even mine/process/enrich the majority of the uranium going into our nuclear reactors[173]. The USA came to rely upon countries more recently empowered/lubricated by "cheap" oil, to meet its people's wants with their still-cheap labor and natural resources. Furthermore, over the last four decades global per capita consumption of some commodities has grown eight to twelve times faster than has human population. After this century's inevitable post-peak-oil period, interruptions in the flow of such goods will be profoundly destabilizing to what had once been the USA's uniquely large & prosperous "middle class". They had become accustomed to enjoying extra consumption via a combination of both their own and their government's addictions to deficit spending plus the "off sourcing" enabling low cost expropriation of other nation's/people's share of the Earth's resources[174]. Demographics, the outsourcing of energy-intensive industries plus automation of most of remaining ones has severely impacted the people who had worked in them, thereby generating poverty, insecurity, plus a great deal of angst and political unrest enhances their vulnerability to populistic scapegoating.

Where did today's anti-human mindset come from? Senator Elizabeth Warren introduced a new piece of legislation in August of 2018 called the Accountable Capitalism Act which provides a snapshot of how it happened. Here's the gist of what it says:

For most of our country's history, American corporations balanced their responsibilities to all of their stakeholders – employees, shareholders, communities – in corporate decisions. In the early 1980s, America's

---

[172] Natural thorium's half-life (14 billion years) is about three times the age of the earth meaning that it's just nominally radioactive. However, because it is generally in secular equilibrium with ten interim decay products ("daughters"), any such ore's total radioactivity is about an order of magnitude greater than that of the thorium itself and therefore easily detected. The radioactivity of a typical rare earth ore containing 2.5% thorium is ~1000 Bq/g - high enough to render working with it under the USA's regulatory environment more expensive/troublesome than it's worth.

[173] Since the collapse of Soviet Union more than 19,000 Russian warheads have been dismantled and their high enriched uranium (HEU) diluted to make fuel for U.S. nuclear reactors. During that period more than half the fissile (^{235}U) powering civilian US reactors came from such "reprocessed" nuclear weapons.

[174] The latter has become especially convenient to the USA's political leadership because they could claim that, "*it's them foreigners, not us, that are really polluting the atmosphere*".

biggest companies dedicated less than half of their profits to shareholders and reinvested the rest in the company but then a "new idea took hold: American corporations should focus only upon maximizing returns to their shareholders (i.e., the rich should become richer). Consequently, big American companies currently dedicate ~93% of earnings to shareholders thereby directing trillions of dollars that could have gone to workers and long-term investments in the company itself. Lately, since the easiest way to increase profit margins is via automation, especially when firing people means that you can stop paying their health and insurance costs, that's what's been happening[175]. Many of the USA's middle class people haven't yet come to grips with the whys of their situation and are responding in the same fashion as did Italy and Germany's citizens during the 1920/30s as evidenced by the outcome of its last presidential election; i.e., *blame "outsiders", especially foreigners*[176].

Consequently the USA's middle class is shrinking both in terms of percentage of its population, and qualitatively, in terms of their quality of life. Many are being squeezed by declining real incomes and rising expenses as they increasingly shoulder the inflationary costs of both corporate greed and grossly inefficient, special interest-driven, government(s). The USA moved away from being a "meritocracy". Wealth inequality has recently increased more than during any other time in its history, while economic mobility has shrunk. Roughly 60 percent of America's wealth is inherited and, like most privileges, is heavily skewed to exclusive groups. In 2012, PBS's Bill Moyer pointed out that the real median US household income was under what it was at the end of the '80s and down 9 percent from its peak in 1999, with the biggest part of that decline, 8.3%, occurring during the preceding five years. The median net worth of a family in 2010 was $77,300, compared to $126,400 just three years earlier because the too-big house that their government's policies had encouraged them to over-invest in didn't pay for itself. In 46 states, the poverty rates had increased during the previous five years, and the national poverty rate had been >15% for the fourth year running. More and more families were and are dropping from the ranks of the USA's once proud and confident middle class into "precarity"[177].

---

[175] In an improperly regulated/motivated capitalistic society doing things like that to low-level employees is as inevitable as entropy.

[176] *"Remember, remember always, that all of us, and you and I especially, are descended from immigrants"* Franklin D. Roosevelt.

[177] i.e., low pay, high blackmailability, intermittent incomes, etc., and, of course, the social consequences of poverty combined with relentless cost-of-living inflation.

One of the actors adversely affecting median household incomes is the scarcity of "good" jobs exacerbated by extended periods of unemployment including deliberate government shutdowns. More and more people are being forced to accept "gigs" rather than secure jobs that pay enough for their holders to achieve the same American Dream – real (not pretend[178]) home and car ownership, etc. – that their parents (my generation) had enjoyed.

The picture isn't much better with respect to the cost of the goods and services that they must consume. The Consumer Price Index (CPI) is currently the most relied upon figure for calculating the yearly inflation rates used to determine pay raises and pension benefits. According to Forbes, the USA's Bureau of Labor Statistics (BLS) changed the way it calculates the CPI twenty times over the past 30 years, including new formulas and indices that separated the more volatile food and energy sectors (e.g., the cost of building nuclear reactors) to arrive at a "core inflation rate". Those changes resulted in a significant dissociation between what the government reports as inflation rate, and what the middle class sees when buying the goods and services they need.

In 2014 Forbes declared, that "*The CPI is not a measurement of rising prices, rather it tracks consumer spending patterns that change as prices change. The CPI doesn't consider the falling value of money. If it did, the CPI would look much different.*"

According to the BLS the CPI had gone up 1.6% that year, and had hovered between 1-4% for the preceding five years. However, if the inflation rate were calculated in the same way as it had been circa 1980, it would've averaged over 5% per year. For example, out-of-pocket healthcare costs had nearly doubled in the preceding seven years because our government's increasingly special interest driven policies enable/encourage "health care providers" and drug manufacturers to charge whatever they can get away with[179].

Domestic energy prices have likewise dramatically increased. During the last decade, energy prices had more than doubled as our governments' policies became increasingly ideological and counterintuitive. Energy and food cost inflation disproportionately affects the middle and lower classes. When I first arrived in Idaho Falls 40 years ago electricity was so cheap ($0.008/kWr) that most of its almost-new houses were simultaneously

178 Nothing is really owned until the last payment has been made. The average US homeowner's equity is currently under 50%.

179 When pain or an early death is the alternative, a service's providers have unusual leverage.

poorly insulated and "all electric" with resistive-type heating. Electrical power was cheap because President Roosevelt's new deal dam building campaigns had rendered it so and the Northwest's population growth hadn't yet pushed power demand beyond its publicly built/owned system's ability to satisfy.

ALICE, (the United Way's acronym for **A**sset **L**imited, **I**ncome **C**onstrained, **E**mployed) which represents individuals and families who are working, but unable to afford the basic necessities of housing, food, child care, health care, and transportation), determined that nearly a quarter of a million of Idaho's households – 40 percent – could not afford those needs in 2016 [206]. While most Americans think of the US as being a country of great economic mobility and opportunity, its economic mobility rate is now one of the worst in the developed world. In the US people whose fathers were in the bottom income quartile have a 40% chance of staying in that quartile and only about an 8% chance of making it to the top quartile, which is half of the average probability of moving up and one of the worst probabilities of the countries analyzed. In a country of equal opportunity, that would not be the case [207].

The childhood poverty rate in the US is now 17.5% and has not meaningfully improved for decades. In the US in 2017, around 17% of children lived in food-insecure homes where at least one family member was unable to acquire adequate food. UNICEF reports that the US is worse than average in the percent of children living in a food-insecure household (with the US faring worse than Poland, Greece, and Chile).

This chapter paints a pretty distressing picture of the USA's middle class's situation. Prospects for improvement are currently also pretty poor because the basis of their problems is causally connected with policies emanating from and firmly entrenched within their government at all levels. Because Congress still has the power to make laws and overcome vetoes and citizens can (in principle anyway) still choose who leads them, it's not just due to its current president's (Trump's) mishandling of virtually everything relevant to addressing his country's impending energy, social, and environmental conundrums. The Republican Party is dominated by people determined to restore economic feudalism and the Democratic Party overly politicizes science and hampers progress by championing both over regulation and unrealistic technical proposals. Capitalists typically don't know how to divide the pie well and socialists typically don't know how to grow it. When such economic polarity and poor conditions obtain, the USA's leaders should pull together to reform the system. They are instead becoming progressively more extreme, fighting more rather than less resulting in today's almost total deadlock on many questions [207]. This book's subjects are too important to just ignore or wring our hands about. Everyone must become willing to do the homework required to understand the seriousness and urgency of their situation and properly evaluate the proposed technological fixes.

The person whose 1988 Congressional testimony sparked widespread awareness of both global warming and its causes within the USA, Dr. James Hansen, was critical of Hillary Clinton's intention to put 500,000,000 solar panels on rooftops across the USA: "*You cannot solve the problem without a fundamental change, which means you have to make the price of fossil fuels honest. Subsidizing solar panels will not solve the problem. We have two political parties and neither wants to face reality. Conservatives pretend it's all a hoax, and liberals propose solutions that are non-solutions*[180]."

I'm hoping that the current crop of liberal presidential candidates lobbying for a "green new deal" become willing to take another look at the assumptions underlying their apparent faith in Mark Z. Jacobson *et al.*'s eminently politically correct but technically unrealistic alternative energy schemes. Both psychology and politics can change quickly. As the enormity of the Anthropocene's climate crisis[181] finally sinks in and the hoped-for carbon savings from politically correct renewables aren't realized, a properly implemented nuclear fuel cycle could become the "new green". Protecting the environment and lifting the developing world out of poverty are progressive causes which may cause the millennials and Gen X's to rethink the values that their boomer parents haven't reexamined since they grooved at the Doobie Brothers' "No Nukes concert" forty years ago.

I also hope that the nuclear establishment's ayatollahs become willing to question their assumptions and dictates. Nuclear power should, could, and must become "renewable, not just a temporary bridge to a future powered with "all of the above" except nuclear energy.

Unfortunately another of human nature's quirks is that in today's political environment, bloviating, uber-capitalistic, demagogues can employ the same "populist" tactics that Mussolini and Hitler did to seize command of government and therefore make things worse for everyone except their real "base" (other rich people).

---

[180] Jacobson *et al.* refer to their campaign as a WWS (wind water and solar)-powered "*Solutions Project*". Other people not so heavily invested in its marvelous assumptions consider it a, "*Non Solutions Project*" [208].

[181] The "Anthropocene" is the epoch dating from the commencement of significant human impact on the Earth's geology and ecosystems, including, but not limited to, anthropogenic climate change. A recent study published in the Proceedings of the National Academy of Sciences posited that a "biological annihilation" akin to a sixth mass extinction event is underway as a result of anthropogenic causes [156].

For example, former Washington Post Beijing correspondent, John Pomfret, recently pointed out [209] that our federal government's recent spate of charges against China's leading telecoms company, Huawei, are a smokescreen for its real concern: U.S. security officials are afraid that Huawei's technology will become the global standard for 5G—a development that would have huge implications, both military and economic, to the global balance of power. This is an entirely legitimate fear for which America's leadership has mostly itself to blame. The technology behind 5G is complicated, but not rocket science and depends mostly upon network density. A successful 5G network would require thousands[182] of small "servers" scattered everywhere across the entire country. China has elevated the construction of such networks to the equivalent of the USA's 1960's-era moon landing project, outspending the U.S. by $24 billion since 2015. In other words, China's advantage lies not in scientific genius but political will. That sort of contest favors nations capable of doing the sorts of things at which China has come to excel and the USA no longer bothers to even attempt – public infrastructure building[183].

Fundamentally, it is a question of values. In the 1930s and '40s, the public sentiment was that the nation would be better off if everyone had reasonably comparable electricity and telephone services. As a result, the federal government established a system of loans and grants to ensure universal access to those key utilities. To help, the FCC set up a system to charge businesses and urban customers slightly higher fees to subsidize the higher costs associated with bringing phone lines to rural areas. The question facing the U.S. now, is whether it is willing to commit to providing affordable broadband service equitably to everyone regardless of who they are or where they live –many of us now don't have access to or can't afford even 3G service.

Pulitzer Prize-winning investigative reporters Donald Barlett and James Steele explain in their book, "The Betrayal of the American Dream" that what's been happening to America's middle class isn't inevitable. It's the direct result of government policy and it could be changed by government actions."

The solution to such malaise should be relatively simple, and recognized by everyone from the chairman of the Federal Reserve to the AFL-CIO's

---

182 The reason for this is that a 5G system utilizes much higher frequencies than does a 4G system which renders it more directional, almost "line of sight".

183 For example, every few years almost everyone in the USA has to submit to arbitrary, corporate-driven, and often disruptive changes in their personal computer's word processing and spreadsheet applications (I'm so suffering now).

leadership. In fact, the labor organization perhaps worded it most succinctly in a piece titled, "How do we fix the U.S. economy?" They declared the first step must be "*to put America back to work because high unemployment keeps wages down. Our goal should be 'full employment', meaning everybody who wants to work should be able to find a decent job.*"

Electrifying the USA's residential, industrial, agricultural, and transportation sectors via the development and then implementation of a appropriately scaled and genuinely sustainable nuclear renaissance would provide millions of genuinely good jobs for its citizens – not just service-type gigs.

One of the things stifling US job growth is the overreach of special interest-driven governmental regulations due to the way that the people we've chosen to represent us (Congressperons) have decided to fund political campaigns[184]. In 2013, a U.S. Chamber of Commerce survey showed that 74% of small businesses were positioning themselves to slash hours, lay off workers, or both primarily due to how the USA had gone about implementing its "Affordable Care Act"[185]. Nearly 300 large companies admitted to reducing hours for their employees in order to get below its mandated 32 hour/week threshold. That's the result of just one piece of poorly thought-out, special interest (health insurance, pharmaceutical, and medical industries[186])-driven, legislation.

In 2012, the USA's House (of Representatives) Committee on Oversight and Government Reform published a report illuminating the role that

---

184 For example, the 2010 Republican-pushed/won, "Citizens United" legal battle overturned the McCain–Feingold Act which had prohibited corporations from funding "electioneering communications" and thereby effectively removed limits to the degree that they could help elect business friendly candidates. *"Government by organized money is just as dangerous as Government by organized mob."* Franklin D. Roosevelt

185 Aka "Obamacare" by the people who had insured that the ACA would end up doing that, not by President Obama himself. He would have much preferred a single payer system which would have served everyone equitably and reduce costs, not just shovel more money into the same special interest-driven system.

186 In 1965, the US federal government created Medicare and Medicaid to provide health care services for its elderly and poor. These programs currently provide some degree of health care for 38 percent of its total population. Unfortunately, the loopholes that were deliberately left in that legislation plus the fact that they didn't cover everyone means that the USA's mostly-privatized health care system has become ridiculously expensive, leaving 41 million US citizens underinsured and 28 million with no coverage at all. A single payer ("Medicare for All") system would foster real competition, provide health care to everyone and, by taking price-setting out of the hands of insurance companies and for-profit hospital managers, drastically reduce administrative, drug, and professional care costs. It would also remove a huge burden from employers which, in turn, would encourage job/business growth.

government has played in suppressing job growth. It concluded that, "*Many regulations and legislation – both existing and proposed – exacerbate the uncertainties created by today's volatile economic environment. Virtually every new rule/regulation has an impact on recovery, competitiveness, and job creation.*" President Obama's own Economic Advisory Panel came to the same conclusion, saying that "*regulations are harming businesses and job creation.*" That panel went on to suggest several measures that could be implemented to quell the expansion of such job/economy-destroying rules and regulations.

No other sector of the USA's economy is as much affected (suppressed) by bureaucratic overreach as is nuclear power. The chief impediment being its Nuclear Regulatory Commission which is empowered to determine whether any real or proposed activity involving "special materials" falls under its purview and then charge whoever wants to do anything with them a fee for what it decides that its licensing and oversight activities are worth (see [210]). It is also notoriously slow in responding to requests for such help/advice/permissions, which holds up projects raising overhead costs. That's one of the realities that caused DOE's NE decision makers to try to substitute "modeling" for experimentation which, of course, didn't work out.

Periods of rising middle class income coincide with periods of economic expansion and growth like those prevailing from several years before the beginning of WW II up until about 1970. During that time the USA's tax system prevented the super-rich from hogging the majority of the nation's wealth (top marginal tax rates were >90%) and its government employed that money to embark upon the massive infrastructure-building projects that quickly "made America great"- Hoover Dam and the string of others that tamed the Mississippi and Columbia, the Tennessee Valley Authority, the interstate highway system, etc., etc., while ensuring that it would be cheap/easy for its citizens – especially its military veterans - to buy homes and get low cost educations all the way up through college. During that time, our government won that biggest of all wars (yet) and neither micromanaged the projects that its taxpayers and bond buyers were funding nor imposed efficiency-killing rules and laws. To the contrary, that government (not today's) put people in charge that were willing to insist that its contractors did the job that were being paid to do – not decide for themselves what that job should be[187]. That's one of the reasons why the

---

[187] DOE no longer devises its own technologies, it chooses between the offerings of entrepreneurial contractors. NASA now does the same thing – in both cases it's a tacit admission that the government no longer entrusts itself to do such things. The leadership of

AEC's NRTS in Idaho, was able to design, build, test, and then decommission >50 nuclear reactors and repeatedly recycle the fissile (highly enriched uranium) so-used during its first two decades.

The other reason is that failure has become acceptable (expected) within the DOE's nuclear complex because (almost) everyone within it knows that it's become almost impossible to do anything that's either "risky" or inconsistent with extant political drivers. Consequently, even though DOE now spends more money (about $32 billion/year) and the need for such work is greater than it was six decades ago, it has become incapable of either doing or managing the sorts of projects that the AEC and its contractors quickly accomplished back then.

Mankind's history is littered with examples of once-dominant nations that had overreached with poorly thought-out plans and policies which crippled their economies and eventually led to collapse due to gross inefficiencies. Great Britain's current "Brexit" situation is apt to lead to that conclusion and the USA seems hell-bent upon following it down the same rat hole.

However, as much as both over and stupid regulation must be curbed, there are some things that do need "centralized planning" among which are those upon which the ultimate success of any technological civilization depends – the educational, health care, and energy systems that enable its people to succeed. The real problem is bad rules and regulations, not their number.

Regulation – about the only positive thing I've seen come out of Mr. Trump's federal government so far is that some of its sillier laws, rules, customs, and regulations are being reviewed; e.g., the USA's approach to labeling its radwastes. If such laws, rules, customs, and regulations aren't replaced with equally inane ones (to be determined) that certainly would be a worthwhile achievement. President Trump must encourage research by cutting the regulatory red tape crippling innovation in any/all nuclear fields. A company wastes up to ten years and $100 million dollars in fees to obtain the Nuclear Regulatory Commission (NRC) permits necessary to conduct research into critical advanced nuclear technologies. The deleterious effect of the USA's regulatory burden on its energy innovation and

---

each of the DOE's national laboratories is chosen by its M&O contractor's top management (e.g., Bechtel's CEO), not DOE. Additionally, that contractor's personnel, not DOE's, make most of the decisions at each site based primarily upon how much money can be made rather than how such efforts further that site's mission. The consequence is that a DOE contractor can and usually does, receive almost 100% of its potential "award fee" while failing (again) to accomplish that site's nominal mission – hence INL's and Hanford's interminable reprocessing radwaste boondoggling.

baseload security cannot be understated. Red tape has driven ThorCon and Terrapower's people and investment dollars to research facilities situated in Indonesia and China, respectively. Streamlining the permitting process for advanced nuclear research will keep atomic energy leadership and jobs here in America.

However, it also seems that most of the people Mr. Trump has chosen to head his science/technology agencies are more determined to undermine them than anything else; i.e., insist upon throwing babies out with the bath water. Some of his agencies should become more, not less, proscriptive/powerful: for instance, *almost* everyone would be better off if the US were to adopt just one measurement system (metric) and limit the number of screw/bolt head choices that manufacturers can use. Why do we pay for a "NIST" that's not empowered to set standards? Any new house being sold should come complete with blueprints informing their purchasers where its pipes, wires, septic tank etc. are located. At the very least such information should be readily available from the governmental agencies issuing building permits.

As far as the future is concerned, the only way that today's internal combustion engine powered cars could be replaced with electric vehicles is if governments dictate a "standard" battery pack configuration that could be as quickly switched at refueling stations as we can now pump 5 gallons of gas into a tank.

Health – today's tremendously overhead-burdened health care system could be rendered much more efficient, cheaper, and effective (fairer) by simply adopting single-payer "medidental" care for everyone and eliminating the 20% coverage gap currently serving to foster most of today's privatized "senior insurance" scamming. Doing so would permit business owners and their employees to concentrate upon running those businesses – not worry about or compete with each other about "benefits". People in the healthcare sector should provide health care services – not waste most of their time trying to please share holders, lawyers, hospital MBAs, and insurance company CEOs[188].

Making free enterprise work as it should will require reducing bureaucratic overhead thereby allowing individuals to focus upon their businesses not "compliance"[189]. Currently, many US citizens get locked into terrible jobs because they can't risk losing their benefits.

---

[188] My experience with "socialized" medicine came during the two years that I served as an enlisted "night man" at Ft. Ord Army Hospital –we did a good job of taking care of people, felt good about it, and no one had to worry about being sued or paying for anything.

[189] INEL/INEEL/INL's employees were (are?) constantly reminded that "compliance" is their employer's overriding mission.

Leveling the playing field between "small" and large businesses will require cutting business-related red tape, simplifying the tax code, ending loopholes, and reducing the overall tax rate. Currently, in spite of the fact that the USA's large businesses are slightly more heavily regulated, in practice they pay less tax than do small ones because they can afford to out/off source more things and pay for a horde of accountants/lawyers and large political campaign donations[190]. Unions should be encouraged. Sweden has industry-wide unions that negotiate wages and therefore neither has nor needs minimum wage laws. Germany's big businesses have worker councils in which everyone comes together to cordially discuss how things should be run. Japan's auto industry did/does the same thing which is why its cars are so well designed, well made, and affordable.

Education – everyone should be given a fair chance of getting whatever education they're both interested in and capable of absorbing[191] from kindergarten to PhD for free. Civics lessons should become a much more important part of that education. A year or so of public service should also become part of everyone's education. The "draft" used to serve that purpose for young US males not rich or "connected" enough to suffer from bone spurs. Some of those draftees learned skills/trades while so serving that subsequently became their means of livelihood[192]. Today's approach to

---

[190] That's surely the reason why the USA's businessman president refuses to release his personal income tax information.

[191] This means free to anyone ready, willing, and able to learn what must be learned to master worthwhile subjects – no more "parapsychology" majors, social promotions, and having your parents buy you into one of the "best" schools ("... *today, FBI Authorities say its operation, dubbed Varsity Blues, uncovered 33 parents described as a 'catalog of wealth and privilege' had collectively paid $25 million to a college admissions counsel who, who had pleaded guilty and agreed to cooperate in an investigation into the 'widening corruption of elite college admissions*'" [211]). The USA's "best" colleges and universities – even state-supported ones - have been allowed to become too frivolous and too expensive (exclusive). On the other hand, Germans believe that education should not be considered a commercial product, and that free access to higher education ensures economic growth and welfare for everyone.

[192] During the Vietnam War while I was completing the US Army's basic training, its leadership decided that I could best serve the nation by becoming a hospital laboratory technician. The ensuing 3-4 month, on-the-job educational experience at Fort Ord's hospital was the most efficient that I'd ever experienced (I'd been drafted out of graduate school and already had a MS in chemistry). I soon became one of the lab's three "night men" entrusted with handling any of the STAT requests that might come in during the sixteen hours each day when no one else was there. Of course, as soon as my two years were up, I was no longer "qualified" to do such work anywhere else in the USA because each individual state allows its already-licensed civilian med techs to decide who might compete with them (that profession is one of the USA's too-many "closed shops").

school funding encourages/enables the super-rich to send their children to prestigious private schools and colleges while simultaneously doing whatever they can to short-change other peoples' kids. It's not really good for anyone except the people selling such high-end services.

Energy – as I've tried to teach in this book, the amount of useful energy serving each person determines his/her life style: happy/sad, rich/poor, free/entrapped, secure/precarious.... That energy must be "clean", cheap, reliable, genuinely sustainable, and sufficiently abundant to provide everyone with a lifestyle comparable to or better than that of today's average European. That combination of characteristics could realistically be realized only with a properly implemented nuclear renaissance – not with biofuels, windmills, solar panel/towers or more-of-the-same unsustainable nuclear reactors. This means that implementing this book's version of the "Age of Substitutability" represents a way for the next couple of generations of US citizens to do for themselves the same sorts of things that Franklin Roosevelt's "new dealing" empowered their great grandparents to do. DOE's management culture fosters poor team player[193] harassment which stifles innovation and enables the people nominally responsible for making such a renaissance possible to ignore that mission to better serve more immediate political/cultural/financial/personal interests.

[193] In many cases simply asking "why" when unethical and/or unreasonable actions were observed resulted in harassment and shunning. I survived (but did not thrive) for over a decade after I'd started crossing that particular Rubicon [212, 213] because I was careful to initially submit my observations/suggestions up through the proper chain of command while reminding everyone that doing so is exactly what the folks who had given us our "ethics training" told us was the right thing to do. When I first began to wonder about how I should respond to some of the things that were happening, I had the good fortune to stumble upon a book, "The High Priests of Waste", written by the US Federal Government's most successful defense-contracting system's whistle blower [214]. When he was very predictably fired for rocking the US Air Force's boat too much, Mr. Fitzgerald appealed to the Civil Service Commission and subsequently won both reinstatement and promotion to the Senior Executive Service where he spent the next thirty years reminding his bosses that their favorite "pet elephants" ("too big to fail" contractors) were supposed to abide by the same rules that everyone one else had to. Nevertheless, whenever those pet elephants (e.g., Lockheed) were going through lean spells between big defense contacts, their close friends in high places continued to make sure that they could make a few bucks managing things like the USA's national laboratories. For an up-to-date example of how that continues to serve the taxpayer see https://www.rt.com/usa/451098-hanford-nuclear-lockheed-lawsuit/.

10
Suggestions for Improvement

The way that DOE goes about labeling, performing and assigning costs to its activities often serves as "the kiss of death" to perfectly reasonable – often the most reasonable - ways of addressing its technical problems. For example, because DOE chooses to refer to its Hanford waste management boondoggle as a "vitrification" project, in many people's minds, that technology has become almost impossibly expensive/difficult to implement. That's baloney, real glass is too cheap to recycle and melter offgas is easy/cheap to clean up. Similarly, DOE's assertion that it would be billing its YM repository's prospective customers (at least INEEL's) at a rate of ~\$800,000/m^3 also made radwaste "disposal" seem ridiculously expensive. How much should it cost to dig one cubic meter's worth of hole in "soft" rock with a second-hand tunnel boring machine? It has similarly managed to demonize both calcination and incineration.

Energy has been the entire world's biggest "real" business[194] for well over a hundred years and devising a viable substitute for the ways that we've gotten used to generating it could make America great (and rich) again. Consequently, DOE should no longer be entrusted with the management of the USA's nuclear engineering/scientific development efforts because that undertaking is too important to entrust to an organization that can't decide what its mission is ("all of the above"?) or how to go about accomplishing it[195]. Its NE division shouldn't just be reorganized again because much of what it does doesn't really have much to do with developing a

[194] "real" as opposed to service industries like selling 5X marked-up Chinese-made goods, banking, insurance, and advertising.

[195] e.g., "all of the above" is a slogan not a plan. The purpose of a nuclear reactor is to generate power, not to be "small", "idiot proof", "manage" spent LWR fuel, or complement more politically correct (wind, solar, biofuels, etc.) ways of generating electricity.

Darryl Siemer (ed.) Nuclear Power: Policies, Practices, and the Future, (179–184)

sustainable nuclear fuel cycle and most of its leaders have proven themselves to be unwilling to either change or lead[196].

Every assumption/rule/law currently engendering foot-dragging on efforts to address Mankind's key technical issues must be critically examined and, whenever necessary, replaced or eliminated. One such assumption is that nuclear power isn't "renewable". It could be and must become so, which should constitute our federal government's NE R&D top dog's vocation, not continuing to help his/its "industrial partners" develop/sell whatever they feel is more "realistic" over the short haul. If he/she wants to actually solve mankind's long term energy conundrum, he/she must become willing to take his/her employees' blinders off and empower them to do the necessary experimentation. Issues like proliferation are political not technical problems because any sort of reactor must have fissile within it which, in principle, could be "diverted" and somehow misused. Again, we voluntarily accept the risks of traveling in cars and airplanes even though those risks are very real and much greater than are those posed by any sort of realistically managed nuclear fuel cycle. Likewise, nuclear waste management's issues are "pretend" issues that would be easy/cheap to deal with if our leaders & their teams of good-team-playing technical experts really wanted to.

Next, any clean energy source's value should be based upon the degree it could serve mankind's long term needs, not upon how much "capacity" it might occasionally provide. If adding such capacity to an energy suppliers' portfolio doesn't enable replacement of their fossil fuel powered generators, its "value" should approximate zero, not considered especially high due to special subsidies or warm fuzziness.

NNSA should not remain a part of whatever becomes of DOE. It should be moved to defense because ....

- ✓ doing so would put responsibility where it belongs
- ✓ the USA's defense establishment has almost infinitely deep pockets
- ✓ it would cut out at least one layer of bureaucracy
- ✓ most of our military leaders are genuinely patriotic (civics minded), cognizant of technical facts, possess an unambiguous chain of command, and, genuinely concerned about the technical issues apt to impact national security; i.e., not *entirely* politically driven (see [7])

---

196 Reorganizations generally just shuffle management personnel sideways to equivalent pay-slots where they're apt to be equally or more incompetent. In the DOE Complex, some of them end up filling slots supposedly reserved for that site's "Fellow" and "Consulting" engineers/scientists.

A hefty carbon tax should be levied upon any fossil fuel that's apt to become GHG. James Hansen has suggested that most[197] such money should be returned to taxpayers/customers via the IRS rather than given to special interest driven/funded politicians to spend as they wish. I agree that this would probably be the fairest and most effective way to affect the necessary changes.

A U.S. GHG reduction policy should be designed to create a level playing field in which all real and proposed energy production technologies can compete in an open marketplace conditioned by legislated CO_2 emissions goals. A CO_2 ("carbon") price for all carbonaceous fuels without long-term subsidies or other preferential policy treatment would be the most effective way to achieve it.

The NRC should oversee the operation and licensing of existing reactors, not dictate how the laboratory and pilot plant experimentation necessary for reactor development is to be done. Its current rules, regulations, lack/level of technical expertise, mindset, and mode of financing[198] constitute one of the entire industry's most formidable "barriers to science". For example, one of the chief drivers/excuses for DOE's grossly inefficient separations-based reprocessing waste management schemes is a several decade old NRC opinion which stated that HLW from which a "maximal effort" had been made to remove its high components[199], no longer had to be considered HLW. Any rules/opinions dictating actions should specify definite toxin/radionuclide concentration limits, not adjectives.

Whatever organization ends up in charge of developing/implementing a sustainable US nuclear renaissance (I'd recommend a new branch of the military[200]) should eschew the personnel management philosophy reflected throughout the DOE's nuclear Complex because it destroys esprit de corps, inhibits creativity, and fosters cynicism. For instance, at INEL/INEEL/INL although everyone "coming on board" during a specific time

[197] Some of that tax money should fund NE-R&D specifically dedicated to devising a sustainable nuclear fuel cycle.

[198] Among the contextual benefits that China, Korea, and Japan reactor builders enjoy relative to Western countries is that the regulator is paid by the government, not the vendor or developer.

[199] Especially "high" constituents include ^{90}Sr, ^{137}Cs, TRU (especially Pu) and in some cases, ^{99}Tc and ^{129}I.

[200] The USA's two most successful nuclear engineering projects were managed by uniformed (not retired) military officers (Groves and Rickover) who wern't confused about whom they were working for or what their mission was. One caveat I'd add though is that it should become absolutely verboten (result in permanent loss of all benefits, honors, pensions, etc.) for any such person to "retire" and go to work for a contractor having anything to do with that mission either directly or as a lobbyist.

was "guaranteed" a set of work related rules and benefits, every succeeding reorganization and contract change encouraged the new contractor inheriting management of both that Site itself and its employees to chip away at those guarantees and keep the money for itself. The excuse proffered by the Site's management whenever anyone screwed up enough courage to point that out was, "*that's the way that things are going everywhere else in good 'ol USA – if you don't like it, you are more than welcome to quit because there's lots of fresh young college grads out there clamoring for any sort of 'good' job these days*". That's no way to convince the best and brightest of any institution's employees to do their best to help make it succeed.

During the years I worked there most of Chem Plant's employees were viewed as cogs in a machine that implemented procedures. It worked fairly well up until the time that its primary mission changed from reprocessing spent fuel to devising "uniquely cost effective ways" of managing its radwastes. Research can't be proceduralized because it requires creative as well as collaborative talents. If an institution's leadership really wants to address new problems, employees motivated to both learn and grow would be much more valuable than those who just want to fit in.

A good sized chunk of money should be devoted to building a test reactor similar to the one depicted in Figure 26 – the one proposed by the folks managing the EU's EVOL (MSFR) program six years ago [215]. It's a relatively but not trivially small reactor (about one cubic meter core volume and up to about 100 MW thermal – enough to power a small city during tests) which could be used to test/develop almost any kind of "fast" molten salt breeder/isobreeder & its attendant fuel cleanup/reprocessing/waste treatment system.

The goal of the studies performed with it should be to devise something that's simultaneously sustainable (CR>=1), maintainable, and affordable, not just make work for experts doing more of whatever they're already comfortable with[201]. DOE is still not considering building a test reactor capable of guiding any such efforts [168].

APPENDIX XIII goes through a specific example of the sorts of questions that a little test reactor like this could answer.

---

[201] For instance, a great deal of work has recently been done at ORNL to "*assess the feasibility of replacing the conventional uranium oxide (UO_2) fuel of the existing fleet of light water reactors (LWRs) with accident-tolerant fully ceramic microencapsulated (FCM) fuel*" [216]. The fuel in question consists of the NGNP's project's TRISO kernels in which their UO_2 has been converted to uranium nitride (actually $UC_{0.25}Si_{0.75}$) and then embedded in silicon carbide instead of pyrolytic graphite. Such fuel would be difficult/expensive to make, virtually impossible to reprocess, and the reactor itself would be little or no more fuel (uranium) efficient than are today's LWRs.

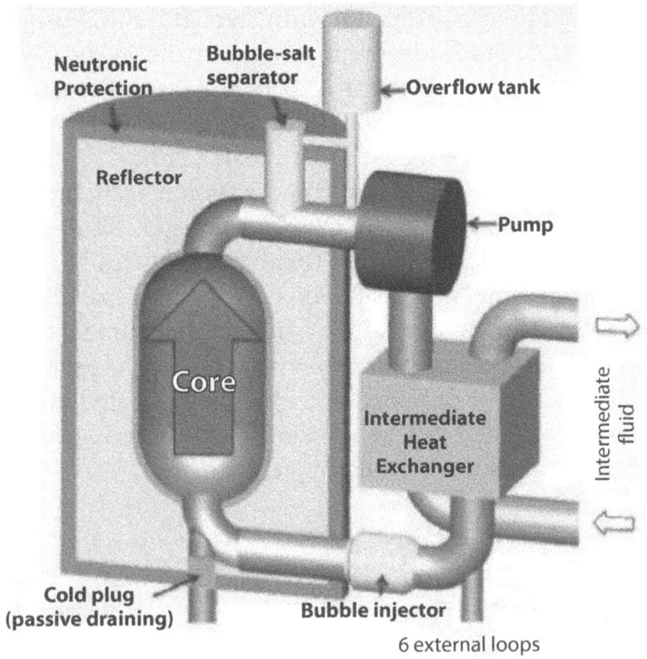

Figure 26 Generic FS MSR test reactor (core volume roughly 1 m^3, heat generation up to 100 MW, its "reflector" volume/tank could either contain a molten bismuth (or tin/lead) neutron reflector or a fertile-containing molten blanket salt.

When one or more of the concepts so investigated meets those criteria, moving it/them to full commercial scale will require additional paradigm shifting because the current incentives for private investments in large scale demonstrations are too weak (the valley of death) and the government's track record in managing large demonstration projects is too poor (the technology pork barrel). Both recent literature [217] and the results of over 500 case studies indicate that policy makers should emphasize/support: prioritizing learning, iteratively upscaling, tolerating minor setbacks, engaging the private sector, knowledge dissemination, and encourage demand pull by removing bureaucratic barriers and giving the concept full credit for its strengths/virtues relative to other renewable energy sources.

A big problem I see with MRS development is that most of the R&D powers-that-be (invariably engineers) don't seem to realize that a genuinely sustainable system would have to be "a chemist's reactor" and therefore don't pay enough attention to its chemical issues – things tend to get "engineered" before the necessary science is done. Most of DOE's radwaste management gurus are similarly blindered which is why the part of its mission that they are responsible for didn't recognize that using a paper-based "kitty litter" liquid adsorbent to sop up decay heat-generating/nitrate-containing radwastes might lead to the "issues" that shut down DOE's $19 billion Waste Isolation Pilot Plant (WIPP) backing-up its cleanup activities elsewhere for over three years [218].

I also feel that the US federal government should again become willing to compete with "free enterprise" in the electricity generation/supply business, especially nuclear-type power. There's nothing really radical about this proposal because the "new deal" dams that enabled the USA to become "the arsenal of democracy" during WW II, were government built, owned, and managed. Nuclear power generation must be done especially "right"[202] meaning that it should be provided by a mission-dedicated institution run like a US Coast guard managed by someone like Admiral Rickover , not by typical US businesspersons or DOE project managers.

[202] In 1971 Alvin Weinberg first used the term "Faustian bargain" to describe nuclear energy: "*We nuclear people have made a Faustian bargain with society. On the one hand we offer—in the catalytic nuclear burner (i.e., the breeder)—an inexhaustible source of energy. Even in the short range, when we use ordinary reactors, we offer energy that is cheaper than energy from fossil fuel. Moreover, this source of energy when properly handled is almost nonpolluting. Whereas fossil-fuel burners emit oxides of carbon, nitrogen, and sulfur... there is no intrinsic reason why nuclear systems must emit any pollutant except heat and traces of radioactivity. But the price that we demand of society for this magical source is both a vigilance from and longevity of our social institutions that we are quite unaccustomed to.*" Expressing such sentiments in a public forum is the reason that Nixon's AEC downsized him.

11

Conclusions

The discovery of nuclear energy represents one of the largest technological leaps that humanity has ever made. It is equaled only by fire, electricity, and heat engines. Those technologies were not just new tools or simply another way of doing the same things. They rewrote the rules at a fundamental level and couldn't be understood using the same terms and rules extant before their invention. Collectively they made today's civilization, industrialization, and mechanized transport possible.

Climate change is currently contributing to a man-made "sixth extinction" comparable in scale to that which killed off the dinosaurs [219]. The most realistic way for humanity to kick its increasingly destructive addiction to fossil fuels would be to implement a nuclear renaissance capable of meeting the energy needs of a bigger, cleaner, and much fairer future world. Because "affordable" uranium could fuel one implemented with today's reactors for under one decade, it must employ breeder-type reactors with close-coupled fuel reprocessing systems. Since the majority of future anthropogenic greenhouse missions will be generated by a developing world that can neither afford to overpay for energy nor undertake a "Manhattan Project" to develop a sustainable nuclear fuel cycle, today's technologically advantaged world has the responsibility to do so. Because molten salt reactors are uniquely well suited to addressing these problems, research seeking to realize their potential should receive top priority. Unfortunately, most of the world's political leaders choose to address causes that are less controversial and more immediate than is the development of a sustainable solution to what many of them apparently still consider only a "future problem".

A properly implemented nuclear renaissance would mean that we would no longer consider energy to be a limiting resource. We would no longer have to conserve it because it could be produced in unlimited quantities by "burning" the almost infinite amount of natural actinides present in common rocks and seawater. We would also not have to destroy the world's remaining wild places and almost every creature living therein just to generate the energy required to keep our civilization going. The natural

Darryl Siemer (ed.) Nuclear Power: Policies, Practices, and the Future, (185–192)

world wouldn't have to do anything beyond serving as a passive heat sink for reactor cooling which service would heat our environment far less than does today's fossil fueled approach to energy generation.

Where it could take us is beyond our imaginations. With that comes backlash. For one thing, most people just don't understand the magnitude of the changes that must be made. Futhermore, nuclear power is scary, because it's a form of energy that very few people understand, but what they do know for sure is that it could be harmful if misused.

Change is always tough for grownups to accept. Our great, great grandparents were probably hearing things like…

> *"Electricity? Are you kidding? That's basically just letting lightning into your home. I think the gas lights work just fine. Electricity is unnecessary. Can't it kill you? That deadly force, inside your home? We don't need this. Running those wires is costly too, and so damn expensive to build this all out."*

It's likely that when the use of fire was first considered by a tribe of our pre-sapient homo ancestors, some of them didn't want it. All they knew is that they'd heard about how fire had destroyed a forest and that someone had been burned when they'd tried to "divert" some of it.

Today's remaining fossil fuels constitute a very much finite resource and Mother Nature won't replace them within a time scale relevant to our species. Since most of the "easy" such fuel has already been mined, they are now and will become more expensive as time passes. Unless something very much like the scenario described in this book comes to pass, people in poor counties will remain poor and the "middle class" of the richer ones will inevitably riot over cost of living increases (especially fuel costs) as is currently happening in France, Mexico, Venezuela, Zimbabwe, Haiti, and Jamaica. Both the obtaining and burning of fossil fuels pollutes the world that we all must share. Because those resources also represent the most efficient source of the raw materials necessary to make the plastics, lubricants, insulation, fabrics etc., that render life pleasant in first world countries, we must quit wasting them just to generate energy services. Wind and solar power backed up with plenty of batteries is suitable for niche applications like powering the "cabins" back in the wilderness that the super-rich occasionally helicopter-in to visit. They are unsuitable (too unreliable) for powering the factories, ships, trains, cars, trucks, farm and mining machinery, etc., that have brought us the benefits of today's technological civilization which is why we must come up with something that can. Today's nuclear fuel cycle isn't scalable (fuel limited) which is why I'm trying to convince folks that we must develop/implement a genuinely suitable replacement and that doing so is possible.

The USA's millennial generation and their successors are currently experiencing economic uncertainties/anxieties similar to those which haunted their great-grandparents during the 1930s. The reasons for their distress are basically the same foibles of "human nature". We/they can no longer afford to allow the tax-supported institutions and experts charged with the responsibility of solving our technical problems to "grow their businesses" by consistently choosing to discharge their responsibilities in grossly inefficient and self-serving ways. DOE's reprocessing wastes could be properly disposed of quickly and cheaply, which feat would demonstrate that a nuclear renaissance's "waste issues" do not constitute another excuse for not implementing one. The USA's recent failures in most things "nuclear" other than providing our military's leaders with thousands of bombs and generating *some* cheap, clean, electricity for its citizens, aren't "disasters" because nobody's been physically injured by them or is even apt to be. However, those failures/boondoggles continue to generate easily documented and hard to refute arguments for why a US "*nuclear renaissance would be both environmentally impactful and prohibitively expensive*" and thereby serve to prevent the USA from doing what must be done to start moving in the right direction again. Those failures also reflect negatively upon the institutions and people responsible for doing the necessary work.

Frankly, it's hard for me to be very optimistic about the future. After all, we now live in a country in which one of our President's most heavily lithified cronies apparently felt that it'd be a good idea to post a "GOFUNDME" request with a set of crosshairs imposed upon a Federal District Judge's picture to raise money for his defense in *her* criminal court. It would be nice to believe that such things couldn't possibly happen but they do.

> "After fifty-five years as a practicing US scientist[203] '*I'm seeing an increasing rejection of science by my fellow U.S. citizens.. Climate change is not only real, it is beginning to displace people around the globe and will almost certainly contribute significantly to increasing conflict over loss of habitable land and dwindling resources. Instead of facing this growing crisis we choose to ignore the science, which is seen as a threat by industries including oil companies. Pay attention to what is happening worldwide: insect and animal species and numbers are dropping precipitously, heat waves and severe weather events are increasing,*

---

203 Most of this paragraph is a comment by "@California Scientist" in response to a New York Times article entitled, "*White House Panel Will Study Whether Climate Change Is a National Security Threat. It Includes a Climate Denialist*". Since I couldn't say it any better, I won't try: I've started it off a bit differently because "@California Scientist" is apparently about a decade younger than I am.

> *the Arctic and Antarctic are heating and melting much faster than predicted. And how do we respond here in the U.S. We elect a president and administration who do not believe in science. It is hard to be optimistic but we have no choice but to advocate for policies and practices based on our best science, and to hope that, somehow, sanity prevails.*" [220]

The good news is that a sizable number of relatively (to me anyway) brilliant young people ([164] and the Weinberg http://Foundation.org), several environmental groups, and some especially distinguished senior scientists have also taken up Goeller and Weinberg's cause [221, 222]. In April 2015, nineteen prominent environmental scholars released "An Ecomodernist Manifesto" representing a declaration of principles for new environmentalism. Its summary says: "*We offer this statement in the belief that both human prosperity and an ecologically vibrant planet are not only possible, but also inseparable. By committing to the real processes, already underway, that have begun to decouple human well-being from environmental destruction, we believe that such a future might be achieved. As such, we embrace an optimistic view toward human capacities and the future.*" [223].

Even better is the fact that many of the young and not-so-young people recently elected to the US Congress realize that the system they're become a part of is badly flawed and that "radical" changes akin to those occurring during the Roosevelt administrations are overdue [224]. Messages like the one that Alex Canerra just sent to them [225] have a reasonably good chance of actually getting through to people like them.

The best news is that there are several especially promising reactor/fuel cycle concepts for the world's serious NE R&D experts to investigate. If I had a say in such things, I'd recommend that a great deal of attention be paid to developing something along the lines of the unmoderated, two fluid, thorium-fueled MSRs studied by ORNL almost a lifetime ago [116] – the same work that inspired this book's tube-in-shell reactor concept (Figure 16). The reason for this is that any such reactor would be cheap, compact, modular, extremely powerful, and require so little startup fissile that enough of them could be built to fully electrify the whole world within three or four decades.

In terms of which country is most likely to do the work required to actually implement my scheme, I'd bet upon China. It's steadily firing up state-of-the-art light water reactors, its new fast reactor went to full power before Christmas (2014), it has begun construction of a new high-temperature gas-cooled nuclear reactor (Chinese HTR) in April (2015) [226]. Most importantly, it has also apparently decided to commit 22 billion yuan ($3.3 billion) to developing the technology abandoned by the US 47 years ago, molten salt reactors. This effort will no doubt be aided by its recent decision (25 March 2019) to collaborate with France – a move that is likely be welcomed by the folks responsible for developing/studying the MSFR. To help address its need for much cleaner electrical power while its MSRs are being researched, China is relying upon 27 conventional nuclear reactors plus 29 Generation III+ (solid fuel) LWRs currently under construction. It also intends to build an additional fifty-seven more estimated to generate a total of 150 GW_e by 2030.

Its chief competitors in the "advanced reactor" business are apt to be Russia and South Korea. Neither have let themselves become dominated (or led) by know-nothings and fear mongers and both build and export affordable nuclear reactors. Russia currently leads in the development of breeding-capable reactors but it is reasonable to expect that when China's "collectivized" mind has decided how best to go about building its own, it'll quickly overwhelm its competition in that arena as it has in every other it has entered[204].

China isn't apt to be leading that particular campaign because it has more experience with nuclear power or can operate reactors more efficiently and safely than do we here in the West. It'll be leading because it's become extremely rich[205] and its political leadership demonstrates both the will and foresight required to pursue activities apt to serve their county's best interests over the long haul. China's leaders seem to accept the fact that it's their responsibility to provide a safe, comfortable, and clean world for their descendants. They also choose to believe in scientific consensus and that both intra and international cooperation is beneficial[206]. Consequently since circa 1978 China's leaders have been demonstrating that it is possible to incentivize their country's entrepreneurs to do the things required to address China's (and the world's) "technical" problems without ceding control/choice of those efforts' ultimate goals. It's been done by lending state support to targeted industries and technologies, particularly those infrastructure-related activities required to address issues like those discussed in this book. We should also remember that China is already heavily

---

204 India has expressed a great deal of interest in developing a thorium breeder reactor-based, nuclear fuel cycle. However, in my opinion, it is unlikely to overtake France, Russia, South Korea or China. It will however continue to be a good customer of countries able to build affordable nuclear reactors – especially if they can "burn" thorium

205 In any real measure of gross national product (GDP), China is already about twice the "size" of the USA. We're still considered "bigger" because we charge each other more for the services that represent ~ 77% of our GDP. GDP "counts" the movement of money, not its value

206 "*Competition has been shown to be useful up to a certain point and no further, but cooperation, which is the thing we must strive for today, begins where competition leaves off.*" — Franklin D. Roosevelt

involved in African development as a part of its 65-nation "Belt and Road" initiative. The opinion of some Western World politicians that such activity is necessarily "greedy and evil" is wrong-headed. It is a good example of "*help-your-neighbor globalization, not colonization, and African agency, not Chinese rapacity*" [227].

Unfortunately, while this little book's clean, green, rational and much more egalitarian future is likely possible, I don't consider it to be a certainty because it would require that technical-based reasoning and border-ignoring philanthropy, not "human nature", will determine what actually happens. Collectively, Homo sapiens including the majority of good team-playing scientists and engineers seem to "think" more like 50 crabs in a gunny sack than do individuals like James Hansen or Albert Einstein. During the next few decades human nature could "trump" rational behavior in China[207] as it has elsewhere. That is the most compelling argument anyone can make to me about why mankind should not pursue a nuclear solution to the future's otherwise inevitable energy-related social, environmental, and economic issues.

There are many people who fear the effects of overpopulation upon the environment and, consequently, think that other people should be left to starve, or freeze, or die of overwork, or…etc., in order to cut their numbers. Overpopulation remains an issue because the world's approach to implementing its leadership's business models severely impacts the natural world while leaving many people poor, ignorant, desperate, and miserable. These are facts we need to confront, not ignore.

If we want to utter the words "brotherhood", "equality", or "compassion" with any semblance of honesty, we must change our business models, the first step of which should be to provide everyone with abundant, cheap, clean, and reliable nuclear energy. If that comes to pass, there won't be a "population problem", global warming will abate, the rivers will run free again, and we'll stop converting the world's last truly natural regions (e.g., the Amazon) into palm oil plantations and soybean farms. Doing so might even bring an end to the Anthropocene's ongoing "Sixth Extinction" [219].

If we refuse to do such things, or too worried about our economy, or think we should only be obligated to attend to our own needs, or willing to do nothing and thereby sentence other people to death and destitution, then the least we can do is to say that that's what we choose to do – not continue to demonize the technology that could best address the technical issues responsible for most of our problems.

---

[207] After all, China's people are still the primary consumers of rhino horns, pangolin parts, bear gall bladders, shark fins, elephant ivory, etc.

In mid-2017, this old but still fairly young-at-heart cynic answered this QUORA question, "*Will nuclear power make a major comeback in the near future in the US?*" with,

> *"I doubt it. The USA has too many cultural barriers in place to do anything that would primarily benefit its own citizens (as opposed to powerful special interests) in the near future (e.g., by 2050). Developing/implementing a nuclear renaissance capable of addressing the consequences of its addiction to fossil fuels will require far more paradigm shifting than our leadership seems capable of."*

I hope that a future generation of US scientists, engineers, and politicians will do what's necessary to prove my QUORA answer wrong. If we become willing to face real threats, overcome unfounded fears, perform the necessary research and then act upon our findings, we could quickly address the future's challenges and provide our descendants with a safer, happier, more equitable, and much less negatively human-impacted world than the one we're living in now. Doing so would obviate the otherwise likely possibility that our descendants will be struggling to survive in Mad Max's world.

References

1. ESA, https://esa.un.org/unpd/wpp/publ, 2015.
2. Trump, http://time.com/5100328/shithole-countries-trump-reactions/, 2018.
3. UNDP, http://hdr.undp.org/en/content/h, 2018.
4. GINI, https://www.cia.gov/library/publications/the-world-factbook/rankorder/2172rank.html, 2018.
5. Dasgupta, P., "World Poverty: Causes and Pathways", Plenary Lecture delivered at the World Bank's Annual Bank Conference on Development Economics (ABCDE), Bangalore, 21-22 May 2003. Subsequently published in B. Pleskovic, and N.H. Stern, eds., World Bank Conference on Development Economics, 2003 (Washington, DC: World Bank), 2004. http://www.econ.cam.ac.uk/people..., 2003.
6. Goldstone, J.A., The New Population Bomb: The Four MegatrendsThat Will Change the World. *Foreign Affairs (Council on Foreign Relations)*, 89, 1, 31–43, doi: 10.2307/20699781, 2010.
7. CNA. Military Advisory Board, "National Security and the Accelerating Risks of Climate Change", May 2014, http://www.cna.org/sites/default..., 2014.
8. Schumacher, E.F., *Small Is Beautiful: A Study of Economics As If People Mattered*, HarperCollins, HB171.S384 1989, 1973.
9. CATO. https://www.cato.org/blog/prosperity-world-population-growth, 2013.
10. McArthur, J., Good Things Grow in Scaled Packages: Africa's Agricultural Challenge in Historical Context https://www.brookings.edu/.../good-things-grow-in-scaled-packages-africas-agricultur..., 2013.
11. Giampietro, M., *Energy use in agriculture, Encyclopedia of Life Sciences*, MacMillan Publishers, Nature Publishing Group, 15 pages. www.els.net, 2002.
12. Human Power, Human power - Wikipedia, 2018.
13. Sims, R.E.H., Energy-Smart Food For People And Climate, FAO Issue Paper. http://www.fao.org/3/i2454e/i2454e00.pdf, 2011.
14. IPCC, Climate Change 2007: Synthesis Report. Contribution of Working Groups I, II and III to the Fourth Assessment Report of the Intergovernmental Panel on Climate Change [Core Writing Team, Pachauri, R.K and Reisinger, A. (eds).]. Geneva, Switzerland, 104 pp., 2007.
15. Borlaug (accessed) https://en.wikipedia.org/wiki/Norman_Borlaug, 2019.

16. Hong, L.-C., "Super El Niño", ISBN 978-981-10-0527-5 https://www.springer.com/us/book...), 2016.
17. Hansen, J., "Climate Change in a nutshell: The Gathering Storm", http://www.columbia.edu/~jeh1/mailings/2018/20181206_Nutshell.pdf, 2018.
18. Stern, R.J., United States cost of military force projection in the Persian Gulf, 1976–200. *Energy Policy*, doi: 10.1016/j.enpol.2010.01.013, 2010.
19. Foreignpolicy, http://www.foreignpolicy.com/art..., 2013.
20. BP, https://www.bp.com/en/global/corporate/energy-economics/energy-outlook.html, 2018.
21. Jackson, R.B., Le Quéré, C., Andrew, R.M., Canadell, J.G., Korsbakken, J.I., Liu, Z., Peters, G.P., Zheng, B., Global Energy Growth Is Outpacing Decarbonization. *Environ. Res. Lett.*, doi: 10.1088/1748-9326/af303, 2018.
22. Yvon-Durocher, G. *et al.*, Methane fluxes show consistent temperature dependence across microbial to ecosystem scales. *Nature*, 507, 7493, 488, doi: 10.1038/nature13164, 2014.
23. Hansen, J., Sato, M., Kharecha, P., Beerling, D., Berner, R., Masson-Delmotte, V., Pagani, M., Raymo, M., Royer, D.L., Zachos, J.C., Target Atmospheric CO_2: Where Should Humanity Aim? *Open Atmos. Sci. J.*, 2, 217–231, 2008. arXiv:0804.1126. Bibcode:OASJ....2.217H. doi:10.2174/187428230080201021 7, 2008.
24. Hansen, J. *et al.*, Ice melt, sea level rise and superstorms: Evidence from paleoclimate data, climate modeling, and modern observations that 2°C global warming could be dangerous. *Atmos. Chem. Phys.*, 16, 3761–3812 www.atmos-chem-phys.net/16/3761/2016/doi:10.5194/acp-16-3761-2016, 2016.
25. LLNL, Lawrence Livermore National Laboratory's annual energy flow charts (its Sanky-diagrams, see https://flowcharts.llnl.gov/commodities/energy/) are based upon DOE/EIA figures which consider both data and assumptions regarding the efficiency with which relevant "sectors" (residential, commercial, industrial, and transportation) convert primary energy to useful "energy services". Recent assumptions have varied more than the data, 2018.
26. Wackernagel, M. and Rees, W., New Society, Gabriola Island, BC, 1996.
27. Nextbigfuture, https://www.nextbigfuture.com/2012/01/united-states-has-3-to-5-trillion.html, 2012.
28. BP, edition, Energy Outlook https://www.bp.com/content/dam/bp-country/fr_ch/PDF, 2017.
29. Hubbert, M., Nuclear energy and the fossil fuels, Publication No. 95, Shell Development Company, Exploration and Production Research Division, Houston, Texas. http://www.hubbertpeak.com/hubbert/1956/1956), 1956.
30. World Nuclear, http://www.world-nuclear.org/inf..., 2018.
31. Moore, B. and Braswell, B.H., The lifetime of excess atmospheric carbon dioxide. *Global Biogeochem. Cycles*, 8, 1, 23–38, doi: 10.1029/93GB03392, 1994.
32. Martin, R., "The dubious promise of bioenergy plus carbon capture", MIT Technology Review, Bioenergy Plus Carbon Capture Sounds Promising,

Looks Dubious, https://www.technologyreview.com/s/544736/the-dubious-promise-of-bioenergy-plus-carbon-capture/, 2016.
33. CSS (accessed) https://en.wikipedia.org/wiki/Ca..., 2018.
34. Statista, Global grain production, 2017 | Statistic, 2017.
35. Wood, "Wood economy" – Wikipedia, 2018.
36. Lal, R., Sequestration of atmospheric CO2 in global carbon pools, Energy & Environmental Science, July issue, DOI: 10.1039/B809492F, 2008.
37. House K., Z., House, C.H., Schrag, D.P., Aziz, M.J., Electrochemical acceleration of chemical weathering as an energetically feasible approach to mitigating anthropogenic climate change. *Environ. Sci. Technol.*, 41, 24, 8464–8470, 2007.
38. Licht, S., STEP: A solar chemical process to end anthropogenic global warming. *J. Phys. Chem. C*, 113, 16283–16292, Also see http://phys.org/news/2010-07-solar-powered-decrease-carbon-dioxide-pre-industrial.html, 2009.
39. NCA4, *Fourth National Climate Assessment*, see both https://nca2018.globalchange.gov/and https://science2017.globalchange..., 2018.
40. Ojuso, J.O., Data Bank: The iso-radiation map of Nigeria. *Solar Wind Technol.*, 7, 563–575, 1990.
41. World Nuclear, http://www.world-nuclear.org/information-library/country-profiles/countries-o-s/south-africa.aspx, 2019.
42. Clark, M. and Tilman, D., Comparative analysis of environmental impacts of agricultural production systems, agricultural input efficiency, and food choice. *Environ. Res. Lett.*, 12, 6, http://iopscience.iop.org/articl..., 2017.
43. Lal, R., Improving Soil Health and Human Protein Nutrition by Pulses-Based Cropping Systems, Advances in Agronomy (Article in press). *Adv. Agron.*, Elsevier Inc., ISSN 0065-2113, http://dx.doi.org/10.1016/bs.agr..., 2017.
44. Van Huis, A., Potential of Insects as Food and Feed in Assuring Food Security. *Annu. Rev. Entomol.*, 563–582, doi: 10.1146, 2012.
45. Phalan, B., Onial, M., Balmford, A., Green., R.E., Reconciling food production and biodiversity conservation: Land sharing and land sparing compared. *Science*, 333, 1289–1291, https://www.researchgate.net/publication/51610716_Reconciling_Food_Production_and_Biodiversity_Conservation_Land_Sharing_and_Land_Sparing_Compared, 2011.
46. FAO, NEW PARTNERSHIP FOR AFRICA'S DEVELOPMENT (NEPAD), Comprehensive Africa Agriculture Development Programme, November, 2002 http://www.fao.org/3/y6831e/y6831e-03.htm, 2002.
47. Pimentel, D., Houser, J., Priess, E., White, O., Fang, H., Mesnick, L., Barsky, T., Tariche, S., Schreck, J., Alpert, S., Water Resources: Agriculture, the Environment, and Society. *BioScience*, 47, 2, 97–106, https://doi.org/10.2307/1313020, 1997.
48. Smil, V., *Energy in nature and society – general energetic of complex systems*, MIT Press, Cambridge, MA, 512 pp., 2008.

49. Tomlinson, J., Nigerian briefing: How engineers can help secure a sustainable economy. *Proc. Inst. Civ. Eng. Energy*, https://doi.org/10.1680/jener.17..., 2018.
50. Israel (accessed) https://en.wikipedia.org/wiki/Israel, 2018.
51. Jewish, http://www.jewishvirtuallibrary..., 2016.
52. Demographics, (accessed) https://en.wikipedia.org/wiki/De..., 2018.
53. RO (accessed) https://en.wikipedia.org/wiki/Re..., 2018.
54. Hussain, A. and Abolaban, F., Nuclear Desalination: A Viable Option for Producing Fresh Water- Feasibility and Techno-Economic Studies. *Life Sci. J.*, 11, 1, 301–307, ISSN:1097-8135, 2014.
55. Desalination (accessed) Desalination - Wikipedia, 2018.
56. PSU. (accessed) Nitrogen Fertilization of Corn, 2018.
57. Frattini, D., Cinti, G., Bidini, G., Desideri, U., Cioffi, R., Jannelli, E., A system approach in energy evaluation of different renewable energies sources integration in ammonia production plants. *Renewable Energy, Elsevier*, 99, C, 472–482, https://www.researchgate.net/publication/305845407_A_system_approach_in_energy_evaluation_of_different_renewable_energies_sources_integration_in_ammonia_production_plants, 2016.
58. Moulton, K.L., West, J., Berner, R.A., Solute flux and mineral mass balance approaches to the quantification of plant effects on silicate weathering. *Am. J. Sci.*, 300, 7, 539–570, 2000.
59. Beerling, D.J., Leake, J.R., Long, S.P., Scholes, J.D., Nelson, T.J. *et al.*, Farming with crops and rocks to address global climate, food and soil security. *Nat. Plants*, 4, 138–147, (also see http://lc3m.org/news/), 2018.
60. Campbell, N.S., PhD thesis University of Glasgow, The use of rockdust and composted materials as soil fertility amendments. http://theses.gla.ac.uk/617/, 2009.
61. Priyono, J. and Gilkes, R.J., Dissolution of milled-silicate rock fertilizers in the soil. *Aust. J. Soil Res.*, 42, 4, doi: 10.1071/SR03138, 2004.
62. Leeman, W.P., Olivine tholeiitic basalts of the Snake River Plain, Idaho, in: *Idaho Bureau of Mines and Geology Bulletin*, vol. 26, B. Bonnichsen and R.M. Breckenridge (Eds.), pp. 181–191 Cenozoic Geology of Idaho, (also characterizes other US basalts), http://geology.isu.edu/Digital_G..., 1982.
63. Siemer, D., Silicate Weathering to Mitigate Climate Change, in: *Advances in Soil Science Soil and Climate*, R. Lal and B.A. Stewart (Eds.), pp. 249–265, Taylor and Francis, 2019.
64. Jankovic, A., Variables affecting the fine grinding of minerals using stirred mills. *Miner. Eng.*, 16, 4, 337–345, 2003.
65. Energy in Africa, Energy in Africa – Wikipedia, 2018.
66. Brook, B.W., Blees, T., Wigley, M.L., Hong, S., Silver Buckshot or Bullet: Is a Future 'Energy Mix' Necessary? *Sustainability*, 10, 2, 302, doi: 10.3390/su10020302. https://www.researchgate.net/pub..., 2018.

67. Zapata, F. and Roy, R.N. (Eds.), « The use of phosphate rocks for sustainable agriculture », (a joint publication of the AO Land and Water Development Division and the International Atomic Energy Agency) Food and Agriculture Organization of the United Nations, Rome, 2004.
68. Taylor, L.L., Beerling, D.J., Quegan, S., Banwart, S.A., Simulating carbon capture by enhanced weathering with croplands: An overview of key processes highlighting areas of future model development. *Biol. Lett.*, 13, 4, AprPMC5414688, https://www.ncbi.nlm.nih.gov/pmc…, 2017.
69. Hardjito, D. and Rangan, B.V., "Development and Properties of Low-Calcium Fly Ash-Based Geopolymer Concrete", Research Report GC 1, Faculty of Engineering, Curtin University of Technology, Perth, Australia (contact author), 2005.
70. Basalt rebar, http://www.monolithic.org/link-to/basalt-fiber-rebar and http://www.basaltft, 2016.
71. Hansen, J. *et al.*, "Assessing "Dangerous Climate Change": Required Reduction of Carbon Emissions to Protect Young People, Future Generations and Nature", PLOS, Published: December 3, 2013, https://doi.org/10.1371/journal.pone.0081648, 2013.
72. Kramer, M.G., Lajtha, K., Audfenkampe, A., Depth trends of soil organic matter C:N and 15N natural abundance controlled by association with minerals. *Biogeochemistry*, 2017, doi: 10.1007/s10533-017-0378-x, 2017.
73. Hartmann, J., West, A.J., Renforth, P., Köhler, P., De La Rocha, C.L. *et al.*, Enhanced chemical weathering as a geoengineering strategy to reduce atmospheric carbon, supply nutrients, and mitigate ocean acidification. *Rev. Geophys.*, 51, https://www.researchgate.net/pub…, 2013.
74. Extension, https://www.extension.iastate.ed)…, 2002.
75. Fortress, https://fortress.wa.gov/ecy/publ…, 2011.
76. Orr, J.C. *et al.*, Anthropogenic ocean acidification over the twenty-first century and its impact on calcifying organisms. *Nature*, 437, 7059, 681–686, Bibcode:2005Natur.437.681O. doi:10.1038/nature04095, 2005.
77. Berner, R.A., A New Look at the Long-term Carbon Cycle. *GSA Today*, 9, 11, 1–6, (pdf) https://www.geosociety.org/gsatoday/archive/9/11/, 1999.
78. Schuiling, R.D. and Krijgsman, P., Enhanced weathering: An effective and cheap tool to sequester CO_2. *Clim. Change*, 74, 1–3, 349–354, http://www.innovationconcepts.eu/res/literatuurSchuiling/enhanced.pdf, 2006.
79. Gilbert, R. and Perl, A., Transportation-in-the-Post-Carbon-World https://www.scribd.com/document/63609779/TRANSPORTATION-Transportation-in-the-Post-Carbon-World-by-Richard-Gilbert-and-Anthony-Perl (chapter in The Post Carbon Reader: Managing the 21st Century's Sustainability Crises, Richard Heinberg and Daniel Lerch, eds. Watershed Media, 2010). 523 p. For other book excerpts, permission to reprint, and purchasing visit http://www.postcarbonreader.com, 2010.

80. Agecon, *Fuel Prices and Cost of Production*, University of Nebraska–Lincoln, https://agecon.unl.edu/cornhusker-economics/2015/fuel-prices-and-cost-of-production, 2015.
81. Cataluña, R., Kuamoto, P.M., Petzhold, C.L., Caramão, E.B., Machado, M.E., da Silva, R., Using Bio-oil Produced by Biomass Pyrolysis as Diesel Fuel. *Energy Fuels*, 27, 11, 6831–6838, doi: 10.1021/ef401644v, 2013.
82. Agrawal, R., Singh, N.R., Ribeiro, F.H., Delgass, W.N., Sustainable fuel for the transportation sector. *PNAS*, 104, 12, 4828–4833, doi: 10.1073 pnas.0609921104, 2007.
83. Kanga, D.W. and Holbrook, J.H., Use of NH3 fuel to achieve deep greenhouse gas reductions from US transportation. *Energy Rep.*, 1, 164–168, https://doi.org/10.1016/j.egyr.2..., 2015.
84. Lloyd's Register, Low Carbon Pathways 2050, 10/18/2016, 2016.
85. UMAS, Navigating Decarbonisation, 02/20/2017, 2017.
86. Technavio analysis, Global dual fuel engine market 2016-2020, 07/21/2016, https://www.technavio.com/report/global-tools-and-components-dual-fuel-engine-market, 2016.
87. Hewlett, P.C., *Lea's Chemistry of Cement and Concrete*, Fourth Edition, ISBN: 978-0-7506-6256-7, 2004.
88. Conly, S., "Here's why China's one-child policy was a good thing", https://www.bostonglobe.com/opin... (see also Sarah Conly, "One Child: Do We Have a Right to More?", (book) Oxford University Press, 2015), 2015.
89. Cowan, R., Nuclear Power Reactors: A Study in Technological Lock-in. *J. Econ. History*, 50, 3, 541–567, Cambridge University press http://www.jstor.org/stable/2122817, 1990.
90. Redbook, http://www.oecd-nea.org/ndd/pubs)..., 2014.
91. Deffeyes, K.S. and MacGregor, I.D., "World Uranium Resources", Scientific American. p. 66, 1980.
92. Höök, M. and Tang, X., Depletion of fossil fuels and anthropogenic climate change: A review. *Energy Policy*, 52, 797–809, 2013.
93. Ulmer-Scholle, D.S., "Uranium — Where Is It Found?"? https://geoinfo.nmt.edu/resources/uranium/where.html, 2018.
94. Bolgren, D. and Menees, J., *Nuclear Fuel Reprocessing*, www.ou.edu/class/che-design/a-design/projects-2007/Nuclear%20Reprocessing.pdf, 2007.
95. Tamada, M., (slides) Collection of uranium from seawater, 2009.11.5TM/IAEA Vienna, 2009.
96. Regalbuto, C., "Past and Future Efforts to Extract Uranium from Seawater", March 21, 2014 http://large.stanford.edu/course..., 2014.
97. Kim, J., Tsouris, C., Oyola, Y., Janke, C.J., Mayes, R.T., Dai, S. *et al.*, Uptake of uranium from seawater by amidoxime-based polymeric adsorbent: Field experiments, modeling, and updated economic assessment. *Ind. Eng. Chem. Res.*, 53, 14, 6076–6083, 2014.

98. Permana, S., Takaki, N., Sekimoto, H., Breeding Capability and Void Reactivity Analysis of Heavy-Water-Cooled Thorium Reactor. *J. Nucl. Sci. Technol.*, 45, 7, 589–600, https://doi.org/10.1080/18811248.2008.9711457, 2008.
99. Greenspan, E., "Fuel-Self-Sustaining Thorium Boiling Water Reactors," funded by DOE NEUP from 10/2011 to 9/2014. Prof. Vujic is co-PI. Collaborators: University of Michigan (co-PI: Downar), MIT (co-PI: Kazimi), BNL (co-PI: Todosow) see https://www.nuc.berkeley.edu/people/ehud_greenspan, 2014.
100. CANDU, (accessed) http://www.energybc.ca/cache/nuclear/www.candu.org/candu_reactors.html, 2019.
101. Triplett, B., Loewen, E., Dooies, B., PRISM: A Competitive Small Modular Sodium Cooled. *Nucl. Technol.*, 178, 186–200, 2012.
102. Till, C. and Chang, Y., Plentiful Energy – the story of the integral fast reactor, CreateSpace, ISBN: 978-1466384606 (free at http://www.thesciencecouncil.com/pdfs/PlentifulEnergy.pdf), 2011.
103. Holcomb, D.E., Flanagan, G.F., Patton, B.W., Gehin, J.C., Howard, R.L., Harrison, T.J., Fast Spectrum Molten Salt Reactor Options, ORNL TM-201111/105 https://info.ornl.gov/sites/publ…, 2011.
104. Fiorina, C., The Molten Salt Fast Reactor as a Fast-Spectrum Candidate for Thorium Implementation. PhD thesis, University of Milan https://www.politesi.polimi.it/b…, 2013.
105. Uranium 233, (accessed), https://www.nuclear-power.net/nuclear-power-plant/nuclear-fuel/uranium/uranium-233/, 2019.
106. Siemer, D., Why the MSFR is the "Best" GEN IV Reactor. *Energy Sci. Eng.*, 3, 2, 83–97, http://onlinelibrary.wiley.com/d…, 2015.
107. Malmbeck, R., Nourry, C., Ougier, M., Souþek, P., Glatz, J.P., Kato, T., Koyama, T., Advanced Fuel Cycle Options. *Energy Procedia*, 7, 2011, 93–102, (Available online at www.sciencedirect.com), 2011.
108. Siemer, D., Improving the Integral Fast Reactor's Proposed Salt Waste Management System. *Nucl. Technol.*, 178, 3, 341–352, 2012.
109. Southern Company and TerraPower Prep for Testing on Molten Salt Reactor, DOE Office of Nuclear Energy press release, 9 Aug. 2018 Southern Company and TerraPower Prep for Testing on Molten Salt Reactor, 2018.
110. Latkowski, *et al.*, Molten Nuclear Fuel Salts and Related Systems and Methods, United States Patent Application 20160189813 http://www.freepatentsonline.com/y2016/0189813.html, 2016.
111. Moltex, "A Technical Introduction to the MOLTEX Energy Technology Portfolio, January 2018 https://www.moltexenergy.com/learnmore/An_Introduction_Moltex_Energy_Technology_Portfolio.pdf [also YouTube videos, "Moltex Energy SSR Fly Through" (https://www.youtube.com/watch?v=4iRF6pilm3s (~3 minutes)) and, "Stable Salt Reactor Technology Introduction" (https://

www.youtube.com/watch?v=R4GSDRqah-0 (~6 minutes)) present easy-to-visualize descriptions of the concept], 2018.
112. Scott, I., A Practical Molten Salt Fission React, International Patent Publication Number WO2014/128457-A1, https://patents.google.com/patent/WO2014128457A1, 2014.
113. TECDOC 1647, "Progress in radioactive graphite waste management". IAEA, 2010 http://www-pub.iaea.org/MTCD/tc, 2010.
114. Rawls, J., "Implications for Waste Handling of the Multiplier Module", June 29, 2010 http://www.nwtrb.gov/meetings/20…, 2010.
115. LeBlanc, https://energyfromthorium.com/2007/08/23/the-modified-geometry-2-fluid-molten-salt-breeder/ http://www.thoriumenergyalliance.com/downloads/TEAC2%20presentations/TEAC2_DavidLeBlanc_2.pdf, 2007.
116. Alexander, L.G., Carrison, D.A., Roberts, J.T., MacPherson, H.G., NUCLEAR CHARACTERISTICS OF SPHERICAL, HOMOGENEOUS, TWO-REGION, MOLTEN-FLUORIDE-SALT REACTORS,ORNL-2751 Reactors-Power TID-4500 (14th ed). https://www.osti.gov/servlets/purl/4200482, 1959.
117. Scott, C.D. and Carter, W.L., Design study of a Continuous Fluorination-Vacuum Distillation System For Regenerating Fuel and Fertile Streams in a Molten Salt Reactor, ORNL-3791, doi:10.2172/4527528, 1966.
118. Robertson, R.C. *et al.*, Conceptual Design Study Of A Single-Fluid Molten-Salt Breeder Reactor, ORNL 4541, 207 pp. Web. doi:10.2172/4030941 http://moltensalt.org/references/static/downloads/pdf/ORNL-4541.pdf, 1971.
119. Robertson, R., Briggs, R., Smith, O., Bettis, E., Two-fluid molten salt reactor design study (STATUS AS OF JANUARY 1968), ORNL-4528 UC-80 http://moltensalt.org/references/static/downloads/pdf/ORNL-4528.pdf, 1970.
120. MSRE (accessed), https://en.wikipedia.org/wiki/Molten-Salt_Reactor_Experiment, 2018.
121. ICTT (accessed) https://www.theicct.org/sites/default/files/publications/Global-shipping-GHG-emissions-2013-2015_ICCT-Report_17102017_vF.pdf, 2019.
122. Weinberg, A., *The First Nuclear Era: The Life and Times of a Technological Fixer*, AIP Press, New York, ISBN 1-56396-358 (Dr. Weinberg wrote seven other books and 541 scientific papers. He considered "The Physical Theory of Nuclear Chain Reactors, co-written with Eugene Wigner, as *probably my most important contribution to science.*" He was a member of the National Academy of Sciences, National Academy of Engineering, and a recipient of the Atoms for Peace and Enrico Fermi Awards), 1994.
123. Mahaffey, J., "Atomic Accidents", Chap. 6, Pegasus Books, 442 p., 2014.
124. Jacobson, M.Z. and Delucchi, M.A., A plan to power 100 percent of the planet with renewables, Scientific American, Nov. issue, 2009.
125. Jacobson, M.Z., Delucchi, M.A., Zack, A.F., Bauer, Z.A.F., Goodman, S.C., Chapman, W.E. *et al.*, "100% Clean and renewable wind, water, and sunlight (WWS) all sector energy roadmaps for 139 countries of the world", http://web.stanford.edu/group/ef…, 2017.

126. Ramana, M.V. and Mian, Z., One size doesn't fit all: Social priorities and technical conflicts for small modular reactors. *Energy Res. Soc. Sci.*, 2, 115–124, http://citeseerx.ist.psu.edu/viewdoc/download?doi=10.1.1.666.5276, 2014.
127. Cohen, B.L., Breeder Reactors: A Renewable Energy Source. *Am. J. Phys.*, 51, 75, 1983.
128. Lallanilla, M., "Facts About Fracking", https://www.livescience.com/34464-what-is-fracking.html, 2018.
129. Presalt, Pre-salt layer - Wikipedia, 2018.
130. Kharecha, P.A. and Hansen, J.E., Prevented Mortality and Greenhouse Gas Emissions from Historical and Projected Nuclear Power. *Environ. Sci. Technol.*, 47, 9, 4889–4895, doi: 10.1021/es3051197, 2013.
131. Unscear, Sources, Effects and Risks of Ionizing Radation, United Nations Scientific Committee on the Effects of Atomic Radiation http://www.unscear.org/unscear/en/publications/2016.html, 2016.
132. Yergin, D., *THE QUEST: Energy, Security, and the Remaking of the Modern World*, The Penguin Press, 2011.
133. Obama, B., "Remarks by the President on Energy in Lanham, Maryland, IBEW [International Brother-hood of Electrical Workers] Local 26, Lanham, Maryland," February 16, 2010, http://www.whitehouse.gov/the-press-office/remarks-president-energy-lanham-maryland, 2010.
134. Forsberg, C. and Petersen, P., Basis for Fluoride Salt–Cooled High-Temperature Reactors with Nuclear Air-Brayton Combined Cycles and Firebrick Resistance-Heated Energy Storage. *Nucl. Technol.*, 196, 1, 13–33, doi: 10.13182/NT16-28. https://www.tandfonline.com/doi/pdf/10.13182/NT16-287, 2016.
135. Energy subsidies, https://en.wikipedia.org/wiki/Energy_subsidies, 2018.
136. Tverberg, G., (accessed) https://www.energycentral.com/c/ec/researchers-have-been-underestimating-cost-wind-and-solar, 2019.
137. SciAm, "Biofuels Might Hold Back Progress Combating Climate Change: The latest IPCC report notes the risks of devoting cropland to fuel rather than food", Scientific American, Tiffany Stecker, ClimateWire on March 31, 2014.
138. Lazard, Lazard's Levelized Cost of Storage Analysis —Version 3.0. https://www.lazard.com/media/450338/lazard-levelized-cost-of-storage-version-30.pdf, 2017.
139. Berman, A., http://www.artberman.com/art-berman-shale-oil-hyped/, 2017.
140. Lovering, J.R., Yip, A., Nordhaus, T., Historical construction costs of global nuclear power reactors. *Energy Policy*, 91, 371–382, 2016.
141. World Nuclear, http://www.world-nuclear.org/information-library/facts-and-figures/world-nuclear-power-reactors-and-uranium-requireme.aspx, 2019.
142. Bettis, E.S. and Robertson, R.C., The Design and Performance Features of a Single Fluid Molten Salt Breeder Reactor. *Nucl. Appl. Technol.*, 8, 2, (Paywalled - free at http://energyfromthorium.com/pdf), 1970.

143. OMB. https://www.whitehouse.gov/sites…, 2017.
144. Engel, J.R., Conceptual Design Characteristics of a Denatured Molten-Salt Reactor With Once-Through Fueling, ORNL/TM-7207 http://moltensalt.org/references/static/ralphmoir/ORNL-TM-7207.pdf, 1980.
145. Maloney, T., Critique of 100% WWS Plan, free at http://www.timothymaloney.net/PDF_downloads_for_Critique_of_100_WWS_Plan.html, 2018.
146. Ault, T. *et al.*, Lithium Isotope Enrichment: Feasible Domestic Enrichment Alternatives, Department of University of California, Berkeley Report UCBTH-12-0055 May 2012 http://fhr.nuc.berkeley.edu/wp-content/uploads/2014/10/12-005_NE-170_Lithium-Enrichment.pdf, 2012.
147. POGO, http://www.pogo.org/blog/2014/08/20140825-no-moreexcuses-for-failing-to-downblend.html, 2014.
148. Hurst, C. and Fort. Leavenworth, K.S., 66027. https://apps.dtic.mil/dtic/tr/fulltext/u2/a525378.pdf, 2010.
149. Stover, D., (November 22), The Myth of Renewable Energy, Bulletin of the Atomic Scientists, https://thebulletin.org/2011/11/the-myth-of-renewable-energy/, 2011.
150. Mahaffey, J., "Atomic Awakening: A New Look at the History and Future of Nuclear Power, Pegasus Books, 344 p., 2009.
151. Siemer, D.*., NWTRB Spring 2013 Board Meeting - April 16, 2013, Richland, WA, https://www.nwtrb.gov/docs/default-source/meetings/2013/april/siemer.pdf?sfvrsn=5, 2013.
152. Siemer, D.*., Molten Salt Breeder Reactor (MSBR) Waste Management. *Nucl. Technol.*, 185, 1, 100–108, 2014.
153. Winograd, I., Radioactive waste disposal in thick unsaturated zones. *Science*, 212, 4502, 1457–1464, 1981.
154. Picket, J.B. *et al.*, "Vitrification and Privatization Success", WSRC-MS-2000-00305, Rev. 1, 1995.
155. Tomioka, O., Japan tries to chip away at mountain of disused solar panels. *NIKKEA Asian Rev.*, https://asia.nikkei.com/Business/Technology/Japan-tries-to-chip-away-at-mountain-of-disused-solar-panels, 2016.
156. Steffen, W., (plus another 15 coauthors), Trajectories of the Earth System in the Anthropocene. PNAS, DOI: 10.1073/pnas.1810141115, 2018.
157. Abdulla, A., Ford, M.J., Morgan, M.G., Victor, D.G., A retrospective analysis of funding and focus in US advanced fission innovation. *Environ. Res. Lett.*, 12, 2017, 084016, http://iopscience.iop.org/article/10.1088/1748-9326/aa7f10/pdf, 2017.
158. Ford, M.J., Abdullab, A., Morgana, M., Granger, V., David, G., Expert assessments of the state of U.S. advanced fission innovation. *Energy Policy*, 108, 2017, 194–200, https://www.sciencedirect.com/science/article/pii/S0301421517303506, 2017.
159. NAP, *Barriers to Science: Technical Management of the Department of Energy Environmental Remediation Program*, The National Academies Press, https://

www.nap.edu/catalog/10229/barriers-to-science-technical-management-of-the-department-of-energy, 1996.
160. Galvin, Task Force on Alternative Futures for the DOE National Laboratories. https://www2.lbl.gov/LBL-PID/Galvin-Report/Galvin-Report.html, 1995.
161. Bivens, J., The potential macroeconomic benefits from increasing infrastructure investment, https://www.epi.org/publication/the-potential-macroeconomic-benefits-from-increasing-infrastructure-investment/, 2017.
162. Siddiqui, F., "No, the Private Sector Won't Pick Up the Slack for Federal Investments in Energy Innovation", https://medium.com/third-way/no-the-private-sector-wont-pick-up-the-slack-for-federal-investments-in-energy-innovation-4b3c2029670a, 2018.
163. Bryce, R., Gusher of Lies: The Dangerous Delusions of Energy Independence, PublicAffairs 2008, ISBN 978-1-58648-321-0, 2013.
164. Beckers, M., "Science a la Carte and the Cherry picking", Create Space ISBN 9781530880690, 2016.
165. Clack, T.M., Qvist, S.A., Apt, J., Basilian, M., Brandt, A.R., Caldeira, K. *et al.*, Evaluation of a proposal for reliable low-cost grid power with 100% wind, water, and solar", Evaluation of a proposal for reliable low-cost grid power with 100% wind, water, and solar. *PNAS*, 114, 26, 6722–6727, https://doi.org/10.1073/pnas.1610381114, 2017.
166. DOE, http://www.world-nuclear-news.or..., 2018.
167. Perry, https://www.energy.gov/articles/..., 2018.
168. Petti, D., Hill, R., Gehin, J., Gougar, H., Strydom, G., Heidet, F., Kinsey, J., Grandy, C., Qualls, A., *Advanced Demonstration and Test Reactor Options Study, INL/EXT-16-37867*, Revision 3, https://art.inl.gov/ART%20Document%20Library/Advanced%20Demonstration%20and%20Test%20Reactor%20Options%20Study/ADTR_Options_Study_Rev3.pdf, 2017.
169. Rumbaugh, R. and Cohn, N., "Resolving Ambiguity: Costing Nuclear Weapons," Stimson Center, May 2012, http://www.stimson.org/images/up..., 2012.
170. Mecklin, J., Bulletin of the Atomic Scientists Volume 75 Issue 1, January 2, 2019 https://thebulletin.org/2019/01/introduction-the-wasteful-and-dangerous-worldwide-nuclear-modernization-craze/?utm_source=Bulletin%20blast&utm_medium=iContact%20email&utm_campaign=Introduction_01072019, 2019.
171. IAEA, "Fast Reactors and Related Fuel Cycles: Next Generation Nuclear Systems for Sustainable Development (FR17)", Yekaterinburg, Russian Federation, 26-29 June 2017, STI/PUB/1836 (ISBN: 978-92-0-108618-1)260 pages Fast Reactors and Related Fuel Cycles: Next Generation Nuclear Systems for Sustainable Development (FR17), 2018.
172. Gates, What I learned at work this year, https://www.geekwire.com/2018/china-option-fades-bill-gates-urges-u-s-take-lead-nuclear-power-good-planet/, 2018.

173. CATO, https://www.cato.org/blog/multi-billion-dollar-boondoggle-doe, 2018.
174. Judy, S., https://www.enr.com/articles/44513-doe-seeks-to-halt-mox-construction-convert-facility-to-nuclear-weapons-plant, 2018.
175. Day, D.E. and Kim, C.E., "Reaction of INCONEL 690 and 693 in Iron Phosphate Melts: Alternative Glasses for Waste Vitrification," DE-FG02-04ER63831, Project 0010255, University of Missouri–Rolla ~Aug. 2005! https://digital.library.unt.edu/ark:/67531/metadc875342/, 2005.
176. Glagolenko, Yu.V., "Management of High-Level Radioactive Wastes from the Mayak…., Chapter 24 of a NAP press book, "An International Spent Nuclear Fuel Storage Facility: Exploring a Russian Site as a Prototype: Proceedings of an International Workshop", https://www.nap.edu/read/11320/chapter/24, 2005.
177. Cary, A., (Feb1), Hanford cleanup costs triple. And that's the 'best case scenario' in a new report, https://www.tri-cityherald.com/news/local/hanford/article225386510.html, 2019.
178. Rechard, R., "Performance Assessment of the Direct Disposal in Unsaturated Tuff of Spent Nuclear Fuel and High-Level Waste Owned by U. S. Department of Energy", SAND94-2563/1-3, March 1995, 1995.
179. Batt, https://www.deq.idaho.gov/media/550338-1995_Settlement_Agreement.pdf, 1995.
180. EPA, Environmental Protection Agency 40 CFR Parts 63, 261, and 270 [June 19, 1998] Hazardous Waste Combustors; Revised Standards; Final Rule—Part 1:RCRA Comparable Fuel Exclusion; Permit Modifications for Hazardous Waste Combustion Units; Notification of Intent To Comply; Waste Minimization and Pollution Prevention Criteria for Compliance, Final rule https://www.govinfo.gov/content/pkg/FR-1998-06-19/pdf/98-15843.pdf (The immediately relevant part of this CFR has to do with its timing: "*Clean Air Act (CAA) sets a maximum time frame of three years for facility owners or operators to comply with Maximum Achievable Control Technology (MACT) emission final standards are published in the Federal Register.*"), 1998.
181. EIS, "Idaho High-Level Waste & Facilities Disposition", Final Environmental Impact Statement, DOE/EIS-0287, September 2002, 2002.
182. Gentilucci, J.A., Miller, J.E., Schultz, W.W., "Technical Review of the Applicability of the Studsvik, Inc./THOR Process to INEEL SBW". TFA 0101 http://pbadupws.nrc.gov/docs/ML0907/ML090780200.pdf, 2001.
183. Siemer, D., Steam Reformation of Sodium Bearing Waste: Pros & Cons. *Ceram. Trans.*, 168, 81–90, (in a" preceedings" – the paper itself was presented at ACerSoc's 2004 radwaste symposium), 2005.
184. Hazen, Pilot Plant Report for Treating SBW Simulants: Mineralizing Flowsheet, WGI/THOR International, report 28266-RT-002, Revision 0, July 2007, 2007.
185. Verger, P., Aurengo, A., Geoffroy, B., Le Guen, B., Iodine kinetics and effectiveness of stable iodine prophylaxis after intake of radioactive iodine: A review. *Thyroid*, 11, 4, 353–60, 2001.

186. Kerala, http://bravenewclimate.com/2015/01/24/what-can-we-learn-from-kerala/, 2015.
187. Ghiassi-nejad, M., Mortazavi, S.M., Cameron, J.R., Niroomand-rad, A., Karam, P.A., Very high background radiation areas of Ramsar, Iran: Preliminary biological studies. *Health Phys.*, 82, 1, 87–93, https://www.ncbi.nlm.nih.gov/pubmed/11769138, 2002.
188. Gesell, T., (Professor of Health Physics)." Current Implementation of the Graded Decision Guidelines (Phase 2) March 21, 2013, http://fmcidaho.com/wp-content/uploads/Slag-Presentation-by-Dr-Tom-Gesell.pdf, 2013.
189. Henriksen, T., "Radiation and Health", The University of Oslo medical and biophysics group, http://www.efn-usa.org/2013-11-06-07-30-02/item/777-radiation-and-health-2015-thormod-henriksen-and-biophysics-group-at-uio-norway, 2015.
190. Calabrese, E.J., How the US National Academy of Sciences misled the world community on cancer risk assessment: new findings challenge historical foundations of the linear dose response. *Arch. Toxicol.*, doi: 10.1007/s00204-013-1105-6. Available at: http://link. springer.com/article/10.1007/s00204-013-1105-6, 2013.
191. Cuttler, J.M., Remedy for Radiation Fear – Discard the Politicized Science. *CNS Bull.*, 34, 4, https://www.ncbi.nlm.nih.gov/pmc/articles/PMC4036393/, 2014.
192. Scott, B.R. and Tharmalingam, S., The LNT model for cancer induction is not supported by radiobiological data. *Chem. Biol. Interact.*, 301, 34–53, https://doi.org/10.1016/j.cbi.2019.01.013 (one of several papers published in this "special issue" devoted to reexamining the LNT hypothesis), 2019.
193. Ward, P. and Brownlee, D., Rare Earth: Why Complex Life Is Uncommon in the Universe, Copernicus, 338 pp. ISBN 978-0387952895, 2000.
194. Silva, R.A. *et al.*, Global premature mortality due to anthropogenic outdoor air pollution and the contribution of past climate change. *Environ. Res. Lett.*, 8, 3, 034005, Bibcode: 2013ERL....8c4005S. doi:10.1088/1748-9326/8/3/034005, 2013.
195. WHO, https://www.who.int/en/news-room/fact-sheets/detail/ambient-(outdoor)-air-quality-and-health, 2018.
196. Twain, M., "Letters from the Earth", posthumously published collated by Bernard DeVoto and published by Harper & Row in 1962, 1909.
197. Gray, E., The great uranium cartel, (book) McClelland & Stewart, ISBN-10: 0771035373, ISBN-13: 978-0771035371 http://earlegray.com/jailed-just-talking-great-uranium-cartel/, 1982.
198. Sapolsky, R., This is your brain on nationalism The Biology of Us and Them, Foreign Affairs/Mar/Apr 2019 https://www.foreignaffairs.com/articles/2019-02-12/your-brain-nationalism, 2019.
199. Goldstein, J.A. and Qvist, S.A., A Bright Future: How Some Countries Have Solved Climate Change and the Rest Can Follow, PublicAffairs, ISBN13: 9781541724105, ISBN10: 1541724100, 2019.

200. Enerdata, https://www.enerdata.net/publications/daily-energy-news/eskom-will-cut-4-gw-south-africa-power-network.html, 2019.
201. Advocates, https://advocateswest.org/former-idaho-governors-andrus-and-batt-fault-doe-for-hiding-information-about-proposed-commercial-spent-fuel-shipments-to-inl/, 2015.
202. SL-1 (accessed) https://en.wikipedia.org/wiki/SL-1, 2019.
203. Simpson, M., Goff, K., Johnson, S., Bateman, K., Battisti, T., Toews, K., Frank, S., Moschetti, T., O'Holleran, T., A description of the ceramic waste formproduction process from the demonstration phase of the electrometallurgical treatment of EBR-II spent fuel. *Nucl. Technol.*, 134, 263–277, 2001.
204. Simpson, M., Projected Salt Waste Production from a Commercial Pyroprocessing Facility. *Sci. Technol. Nucl. Install.*, Article ID 945858, doi.org/10.1155/2013/945858, 2013.
205. Conca, J., Classifications of nuclear waste need a shorter half-life, https://www.tri-cityherald.com/opinion/opn-columns-blogs/article220376080, 2019.
206. ALICE, https://www.unitedwaysei.org/sites/unitedwaysei.org/files/18UW_ALICE_Report_ID_refresh_Lowres_5.10.18_Final.pdf, 2018.
207. Dalio, R., Why and How Capitalism Needs To Be Reformed, https://economicprinciples.org/Why-and-How-Capitalism-Needs-To-Be-Reformed/, 2019.
208. Beckers, M., "The Non solutions project", Create Space, ISBN: 1537673807, 2017.
209. Pomfret, J., https://www.washingtonpost.com/opinions/2019/02/04/deeper-tech-concern-is-core-us-huawei-spat, 2019.
210. NRC, https://www.govinfo.gov/content/pkg/FR-2019-01-31/pdf/2019-00219.pdf, 2019.
211. NYTIMES, https://www.nytimes.com/2019/03/12/us/college-admissions-cheating-scandal.html, 2019.
212. INEEL, Citizen Advisors Back Chemist's Nuke Waste Plan", page A3, Idaho State Journal (Pocatello ID), Feb. 1, 1997.
213. INEL, Chemist Advises That High Level Waste be Cement, Not Glass", Idaho State Journal (Pocatello), p. A-3, November 20, 1996. [This article appeared the day after I presented my proposal – converting INEEL calcines to a HIP vitrifiable geopolymeric type concrete - to a meeting of INEL's "Site Specific Advisory Board". The meeting was particularly interesting because an attending EPA official challenged DOE's self serving interpretation of EPA radwaste law], 1996.
214. Fitzgerald, E. and Ernest, The High Priests of Waste, W. W. Norton and Company, ISBN 0-393-05359-8 (Buy it)!, 1972.
215. Lucotte, E.M., Introduction to the physics of the molten salt fast reactor, Thorium Energy conference, (ThEC13) CERN, https://cds.cern.ch/record/1625815?ln=pt, 2013.
216. Powers, J.J., Lee, W.J., Venneri, F., Snead, L.L., Jo, C.K., Hwang, D.H. *et al.*, *Fully Ceramic Microencapsulated (FCM) Replacement Fuel for LWRs, ORNL/*

TM-2013/173, https://digital.library.unt.edu/.../1/high_res_d/1087039.pdf, 2013.

217. Nemet, G.F., Zipperer, V., Kraus, M., The valley of death, the technology pork barrel, and public support for large demonstration projects. *Energy Policy*, 119, 154–167, https://doi.org/10.1016/j.enpol.2018.04.008, 2018.
218. WIPP, https://en.wikipedia.org/wiki/Waste_Isolation_Pilot_Plant), 2019.
219. Kolbert, E., *The Sixth Extinction: An Unnatural History*, Henry Holt and Company, 2014.
220. @California Scientist, comment to "White House Panel Will Study Whether Climate Change Is a National Security Threat. It Includes a Climate Denialist", Post by Coral Davenport Feb. 20, 2019, https://www.nytimes.com/2019/02/20/climate/climate-national-security-threat.html?action=-click&module=Top+S, 2019.
221. Pro nuclear, Pro-nuclear movement - Wikipedia, 2018.
222. Pandora's promise, https://www.youtube.com/watch?v=ObcgG9vjUbs, 2013.
223. Ecomodernism, The Breakthrough Institute, 2018.
224. WP, The Green New Deal? A Green New Deal? Whatever it is, 2020 Democrats support it. https://www.washingtonpost.com/politics/the-green-new-deal-a-green-new-deal-whatever-it-is-2020-democrats-support-it/2019/01/30/a75166d4-2340-11e9-ad53-824486280311_story.html?utm_term =.c140c41983f0, 2019.
225. Canerra, (letter to Senator Markey & Representative Ocasio-Cortez dated 3Apr2019) http://energyrealityproject.com/wp-content/uploads/2019/04/MarkeyAOC.pdf, 2019.
226. Forbes, https://www.forbes.com/sites/jam..., 2015.
227. Bräutigam, D., *U.S. politicians get China in Africa all wrong*, https://www.washingtonpost.com/n..., 2018.
228. Staiger, M.D. and Swenson, M.C., Calcined Waste Storage at the Idaho Nuclear Technology and Engineering Center INEEL/EXT-98-00455 Revision 4. https://www.osti.gov/servlets/purl/1033216, 2011.
229. Benedict, M., Pigford, T., Levy, H., *Nuclear Chemical Engineering*, 2nd Ed., McGraw-Hill, 1981 ISBN-10: 0070045313, 1981.
230. Delpech, S., Salanne, M., Simoni, E., Picard, G., Actinides/Lanthanides Separation for the Thorium Molten Salt Reactor, Montpellier, May 2008, https://www.researchgate.net/publication/242151060_ACTINIDES LANTHANIDES_SEPARATION_FOR_THE_THORIUM_MOLTEN_SALT_REACTOR_FUEL_TREATMENT, 2008.
231. Smith F., J., Ferris, L.M., Thompson, C.T., Liquid-Vapor Equilibria in LiF-BeF2 and Li-BF3-ThF4 Systems, ORNL-4415, UC-80 - Reactor Technology, 1969.
232. Greenspan, E., Maximum Fuel Utilization in Fast Reactors Without Chemical Reprocessing, Integrated University Programs, NEUP PROJECT SUMMARY REPORT, Project No. 09-769 https://neup.inl.gov/SiteAssets/Final%20%20Reports/NEUP_Project_No_09-769_Final_Report.pdf, 2012.

233. Dickey, B.R., Wheeler, B.R., Buckham, J.A., High-Level Waste Solidification: Applicability Offluidized-Bed Calcination To Commercial Wastes, https://www.osti.gov/servlets/purl/4276591, 1974.
234. Siemer, D., "Hot Isostatically Pressed Concrete as a Radwaste Form", Advances in Ceramics, Volume 61, Ceramics Transactions, pp. 657-664 (Proceedings of 1995 American Ceramics Society Symposium on Waste Management, Cincinnati, OH, April 29 - May 4, 1995), 1995.
235. Atkinson, H.V. and Davies, S., Fundamental aspects of hot isostatic pressing: An overview. *Metall. Mater. Trans. A*, 31, 12, 2981–3000, 2000-12-01, doi: 10.1007/s11661-000-0078-2. ISSN 1073-5623.
236. Loeding, J.W. *et al.* The Fluidized Bed Calcination of Radioactive Waste", ANL-6322, May 1961, 1961.
237. Petrie, E.C., Letter to E.J. Bailey, Petr-13-65A, "Report on Run 12-Twelve Inch Diametrr Calciner", December 30, 1965.
238. Bonano, E.J., Chu, M.S.Y., Price, L.L., Conrad, S.H., Dickman, P.T. The disposal of orphan wastes using the greater confinement disposal concept, Waste management '91; Tucson, AZ (USA); 24-28 Feb 1991 (slideset at https://www.osti.gov/servlets/purl/1408347), 1991.
239. Baer, T.A., Price, L.L., Gallegos, D.P., Performance assessment of the Greater Confinement Disposal facility on the Nevada Test Site: Comparing the performance of two conceptual site models, Waste management '94; Tucson, AZ (United States); 27 Feb - 3 Mar 1994, SAND–93-2174C (OSTI DE94006195), 1993.
240. Pareizs, J.M., Jantzen, C.M., Lorier, T.H., Durability Testing of Fluidized Bed Steam Reformer (FBSR) Waste Forms for High Sodium Wastes at Hanford and Idaho, WSRC-TR-2005-00102 (U), 2005.
241. Wesson, S.P., Vajo J., J., Ross, S., Determination of specific surface areas of glass filaments by BET and CAEDMON methods. *J. Colloid Interface Sci.*, 94, 2, 552–563, https://doi.org/10.1016/0021-9797(83)90295, 1983.
242. THOR, no date. THOR^{sm} Steam Reforming Process For Hazardous and Radioactive Wastes Technology Report TR- SR02-1, Rev. 1 https://www.nrc.gov/docs/ML1207/ML120730124.pdf.
243. Siemer, D.D., Grutzeck, M.W., Scheetz, B.E., "Cementitious Solidification of Steam Reformed DOE Salt Wastes", ECI Alternative Nuclear Waste Forms, Girdwood, AL January 18-23, 2004, 2004.
244. Soelberg, N.R., Marshall, D.W., Bates, S.O., Taylor, D.D., "Phase-2 THOR Steam Reforming Tests for Sodium-Bearing Waste Treatment, INEEL/EXT-04-01493, Idaho National Engineering and Environmental Laboratory, Idaho Falls, Idaho, January 2004. https://inldigitallibrary.inl.gov/sites/sti/sti/2553196.pdf, 2004.
245. 75 FR 1, "Amended Record of Decision: Idaho High-Level Waste and Facilities Disposition Final Environmental Impact Statement Revised by State 12/21/09," U.S. Federal Register, Department of Energy, pp. 137–140, January 4, 2010], 2010.

246. Alvarez, R.A., Pacala, S.W., Winebrake, J.J., Chameides, W.L., Hamburg, S.P., Greater focus needed on methane leakage from natural gas infrastructure. *PNAS*, 109, 17, 6435–6440, 2012.
247. Barker, R., Idaho trout left high and dry by irrigation ditches - Idaho Statesman, October 18, 2012 12:34 PM https://www.mcclatchydc.com/news/nation-world/national/article24738910.html, 2012.
248. Beneš, O. and Konings, R.J.M., Thermodynamic evaluation of the NaCl–$MgCl_2$ –UCl_3-$PuCl_3$ system. *J. Nucl. Mater.*, 375, 202–208, 2008. https://kundoc.com/pdf-thermodynamic-evaluation-of-the-naclmgcl2ucl3pu-cl3-system-.html, 2008.
249. Bryce, R., Power Hungry: The Myths of "Green" Energy and the Real Fuels of the Future, PublicAffairs, ISBN 978-1-58648-789-8, 2010.
250. Buffet, W., https://www.usnews.com/opinion/b, 2014.
251. Carey, J. *et al.*, Temperature response of soil respiration largely unaltered with experimental warming. *Proc. Natl. Acad. Sci.*, 113, 48, 13797, doi: 10.1073/pnas.1605365113, 2016.
252. Cheap, The Cheapest Places to Live in the World - 2018. https://www.cheapestdestinationsblog.com/2018/12/10/cheapest-places-to-live-2019/, 2019.
253. Cravens, G., *Power to Save The World: The Truth About Nuclear Power*, Knopf, NY, 2007.
254. Dailykos, https://www.dailykos.com/stories/2016/1/27/1476074/-Bird-Deaths-Might-Be-Due, 2016.
255. Duncan, D. and Swanson, E., Organic-Rich Shale of the United States and World Land Areas, Geological Survey Circular 523, 1965.
256. Ebelmen, J.J., Sur les produits de la décomposition desespèces minérales de la famille des silicates. *Anna. Mines*, 7, 3–66, 1845.
257. EFN, (Environmentalists for Nuclear Energy) Frame Page, 2018.
258. Erickson, G., (accessed) "Unintended Consequences: The Lie that killed millions and accelerated climate change, web version is free at http://www.tundracub.com/htmls/unintendedconsequences.html, 2019.
259. Goeller, H.E. and Weinberg, A.M., The Age of Substitutability. *Science*, 191, 4228, 683, (paywalled, original (1974) version available gratis at http://www.osti.gov/scitech/serv), 1976.
260. Heinberg, R. and Lerch, D. (Eds.), The Post Carbon Reader: Managing the 21st Century's Sustainability Crises, 523 p book, Watershed Media, 2010 - 523 p its wel-illustrated chapters are open access at https://www.scribd.com/lists/3724981/The-Post-Carbon-Reader-Managing-the-21st-Century-s-Sustainability-Crises-2010i, 2010.
261. Kelley, C.P., Mohtadi, S., Cane, M.A., Seager, R., Kushnir, Y., Climate change in the Fertile Crescent and implications of the recent Syrian drought. *PNAS*, 112, 11, http://3241-3246www.pnas.org/cgi, 2015.
262. Krepel, J., Comparison of U-Pu and Th-U cycles in MSR: Advanced Nuclear System Group, Paul Scherrer Institut, ThEC2018 conference 29-31. October 2018, Brussels, Belg, 2018.

263. Lane, J.A., MacPherson, H.G., Maslan, F., *Fluid Fuel Reactors (book), Chapter 14*, Addison-Wesley, 1958.
264. Levine, R. and Rubinstein, Y., "Smart and Illicit: Who Becomes an Entrepreneur and Do They Earn More?" NBER Working Paper No. w19276. https://ssrn.com/abstract=2308245, 2013.
265. Lilienthal, D., Atomic Energy: A New Start, ISBN: 0060126175l, 1980.
266. Mackay, D., *Sustainable Energy: Without the Hot Air*, UIT Cambridge, ISBN 0-9544529-3-3, 2009.
267. Martin, R., "Terrapower quietly explores new nuclear reactor strategy", MIT Technology Review. https://www.technologyreview.com/s/542686/terrapower-quietly-explores-new-nuclear-reactor-strategy/Retrieved 2015-11-30, 2015.
268. Martin, R., Ivanpah's Problems Could Signal the End of Concentrated Solar in the U.S.:Canceled solar thermal projects are likely to mean the technology's future is dim in the U.S., so companies are looking overseas.,MIT technology, March 24, 2016, also see https://en.wikipedia.org/wiki/Ivanpah_Solar_Power_Facility, 2016.
269. Mason, R., Tennekes, H., Sánchez-Bayo, F., Jepsen, P.U., Immune Suppression by Neonicotinoid Insecticides at the Root of Global Wildlife Declines. *J. Toxicol. Environ. Immunol.*, 1, 3–12, https://www.boerenlandvogels.nl/sites/default/files/JEIT%20Immune%20Suppression%20pdf_7.pdf, 2013.
270. MIT, *The Future of Nuclear Power: An Interdisciplinary MIT Study*, https://web.mit.edu/nuclearpower/pdf/nuclearpower-full.pdf, 2013.
271. ORNL, http://moltensalt.org/references/static/downloads/pdf/ORNL-3913.pdf, 3913.
272. Orwell, G., *Animal Farm: A Fairy story*, Secker and Warburg, London, England, 112 p., 1945.
273. Pimentel, D., Marklein, A., Toth, M.A., Karpoff, M.N., Oaul, G.S., McCormick, R., Kyriazis, J., Krueger, T., Food versus fuels –environmental and economic costs, doi: 10.1007/s10745-009-9215-8. http://www.stopogm.net/sites/sto…, 2009.
274. QUORA, https://www.quora.com/Why-would-you-support-or-not-support-only-using-nuclear-energy, 2018.
275. Rhodes, R., *Nuclear Renewal: Common Sense about Energy*, Penguin publishing group, ISBN: 0670852074, 1993.
276. Rhodes, R., Energy: A human history, Simon and Schuster, 2018.
277. Riley, B., Joanna, J., DelCul, G., Vienna, J., Contescu, C., Forsberg, C.W., Molten salt reactor waste and effluent management strategies: A review. *Nucl. Eng. Des.*, 345, 2019, 94–109, 2019.
278. Siemer, D., Improving the Integral Fast Reactor's Proposed Salt Waste Management System. *Nucl. Technol.*, 178, 3, 341–352, 2013.
279. Siemer, D., NWTRB Summer 2014 Board Meeting - August 6, 2014, August 6, 2014, Idaho Falls, ID, https://www.nwtrb.gov/docs/default-source/meetings/2014/august/siemer-pc.pdf?sfvrsn=8, 2014.

280. Siemer, D., The Nuclear Option, in: *Advances in Soil Science: Soil Degradation and Restoration in AfricaTaylor and Francis*, Lal, R. and Stewart, B.A. (Eds.), (in press), 2020.
281. Springer, N. and Duchin, F., Feeding nine billion people sustainably: Conserving land and water through shifting diets and changes in technologies", Environ. *Sci. Technol.*, 48, 4444–4451, 2014.
282. Tamada, M., (seminar) "Current status of technology for collection of uranium from seawater", https://www.worldscientific.com/doi/abs/10.1142/9789814327503_0026, 2009.
283. World natural gas, https://seekingalpha.com/article/4196287-world-natural-gas-2018minus-2050-world-energy-annual-report-part-3, 2018.
284. BGR Energy Study, Data and Developments Concerning German and Global Energy Supplies – Summary https://www.bgr.bund.de/EN/Themen/Energie/Produkte/energy_study_2017_summary_en.html, 2017.
285. FAO, FAOSTAT. Food and Agriculture Organization of the United Nations, http://faostat.fao.org/default.aspx, 2011.
286. Abelson, P.H., Energy For Tomorrow (book). ISBN: 0295954132 (Perhaps his most famous work from this time period is an editorial entitled "Enough of Pessimism" ("enough of pessimism, it only leads to paralysis and decay") which became the title of a 100 essay collection published in 1985), 1975.
287. NEI, nuclear costs in context - Nuclear Energy Institute, https://www.nei.org/CorporateSite/media/...and.../nuclear-costs-context-201810.pdf, 2017.
288. NWTRB, Nuclear Waste Technical Review Board, Spring 2013 Hanford meeting (Siemer's "comments") https://www.nwtrb.gov/meetings/past-meetings/spring-2013, 2013.

Appendix I. Reprocessing

Nuclear reactors like automobiles must consume fuel to do useful work. When fossil fuels (typically gasoline, diesel oil, or propane) are burned to power an engine, they are consumed both immediately and almost completely. When that fuel is gone, energy production totally stops. Nuclear reactors are usually incapable of achieving such near complete burn-up because as their fuel (usually fissile uranium) is burned via nuclear fission, a variety of other elements – fission products (FP) – are created and become intimately associated with the remaining fissile. Because they absorb the neutrons that energize fission, such FP accumulations eventually poison the system and stop the reactor, Some fission products may also damage the structural integrity of solid-type fuel assemblies. Even though a good deal of fissile material may remain in such fuels, it can't be burned until separated from neutron-absorbing fission products and possibly compromised cladding via some form of "reprocessing".

Let's look into the history of the nuclear fuel reprocessing facility about which most of this book's examples were derived.

The USA's Atomic Energy Act of 1946 established the civilian Atomic Energy Commission (AEC) to direct atomic energy "toward improving the public welfare, increasing the standard of living, strengthening free competition among private enterprises so far as practicable, and cementing world peace." That Act required the AEC to conduct research in nuclear energy "through its own facilities…." Consequently, in 1949, following a nation-wide search, the AEC established the NRTS in southeastern Idaho, its first field test facility and only dedicated reactor proving ground. That area was chosen because of its isolation, climate, favorable geology, abundant subsurface water, and social-economic factors including local government support, existing infrastructure, and ready availability of manpower, land, and construction materials. The NRTS originally covered 177,000 acres but eventually expanded to cover 569,000 acres (890 square miles) of Idaho's "high desert" in the northeastern end of its Snake River Plain. The Site's name was changed three times, in 1974 to the Idaho National Engineering Laboratory (INEL), again in 1997 to the Idaho National

Engineering and Environmental Laboratory (INEEL), and finally to the Idaho National Laboratory (INL) in 2005. Its original mission was to provide a suitably remote (safe) area where nuclear energy research, development, and testing could be conducted with minimal impact to the public. Initially the AEC planned to construct only five reactors there over a ten-year period. However, rapidly expanding technologies and a nationwide sense of optimism (as opposed to today) that saw no limits to the potential of nuclear energy greatly extended its mission. Eventually, fifty-two, mostly first-of-a-kind, reactors were constructed at the NRTS including Rickover's first pressurized water submarine reactor (PWR) and Untermyer's boiling water reactor (BWR) which between them came to dominate civilian nuclear power worldwide. The NRTS's experimentation regarding reactor safety, design, and applications influenced virtually every other reactor in the world. The first four facilities circa 1950 included the Experimental Breeder Reactor I (EBR-I), the Naval Reactor Facility (NRF), the MaterialsTest Reactor (MTR), and the Idaho Chemical (re)Processing Plant (ICPP).

Succeeding expansion included several additional centers each reflecting a new direction in nuclear research or filling necessary operational functions. From the beginning, political tides and changing world events periodically shifted priorities. Although those influences varied, they fell into two categories: national defense and peaceful applications.

Throughout history, almost all government facilities everywhere have bent to the needs of national defense at one time or another. Conceived as it was in the aftermath of World War II, the initial throes of the Korean War, and the heightening tensions which defined the Cold War, the NRTS exhibited both extremes. For example, its Naval Reactors Facility, devoted itself entirely to projects related to national defense, the most prominent of which was development of the reactors that power most of the U.S. Navy's submarines and aircraft carriers. Other NRTS facilities contributed to that cause less directly. For example, the ICPP's construction schedules were given a boost when shortages of uranium and plutonium threatened to compromise weapons production at other AEC labs. Consequently, the first batch of uranium recovered/recycled by it came from the Hanford Site's plutonium production reactors.

However, not many of its subsequent fuel reprocessing runs had to do with weapons production or even national defense. President Eisenhower's "Atoms for Peace" program encouraged the development of peaceful uses for nuclear energy (electricity generation, medical applications, food irradiation/preservation, etc.) and much of the subsequent work done at the NRTS was directed toward furthering those efforts. One of the most pressing questions at that time was how different materials and components (e.g., fuel assemblies, fuel cladding, and reactor construction materials) of civilian power reactors would perform under intense irradiation. That why the Material Test reactor (MTR), a high flux, water thermalized neutron reactor, was designed to determine, how well the materials being considered for use in those reactors would withstand several years of exposure. This testing was accomplished by bombarding materials ("coupons') with thermalized neutrons. Since the MTR was a very "hot" reactor enough fission products built up within its aluminum-clad highly enriched (bomb grade) uranium fuel assemblies to bring everything to a halt every 17 days or so. Its so depleted fuel assemblies then had to be removed and replaced with fresh ones. Less than 30% of the original ^{235}U in such "spent" fuel was burned per cycle which means that it had to be recovered for recycle. The NRTS's reprocessing facility, Idaho Chemical Processing Plant (ICPP aka "Chem Plant") was designed to be the MTR's support facility and fulfilled that mission for nearly four decades, recovering millions of dollars worth of fissile (^{235}U) that was fabricated into new fuel elements for both it and the Site's other test reactors.

The NRTS's Chem Plant was far more versatile than were those of the federal government's "production" (of bomb grade plutonium and tritium) facilities (primarily Savannah River and Hanford) It could recover the uranium within fuel assemblies clad with virtually any sort of material - aluminum, stainless steel, zirconium, graphite or ceramic composites.

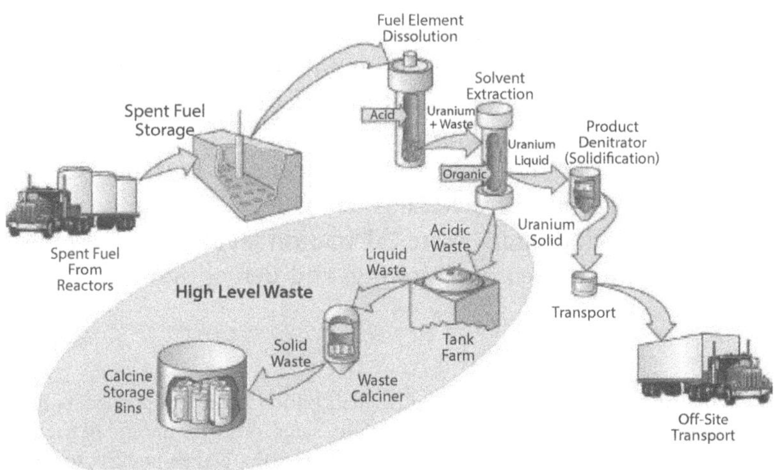

Figure 27 INL's Chem Plant [228].

It also did a better job of managing its reprocessing wastes in that, rather than simply neutralize the acid in them and store the resulting mixture of oxy/hydroxyl sludges plus salt saturated liquids in tanks, about 90% of it was calcined and the resulting more stable/innocuous sand-like dry powder stored in stainless steel/concrete "binsets".

Throughout the years, the Chem Plant recovered both uranium and certain valuable fission products from nearly one hundred different research and power reactors throughout the nation. By the end of the 1970's most of the DOE's test reactors had been shut down ("decommissioned") so it was reconfigured/repurposed so that it could, in principal at least, recover the highly enriched uranium remaining in spent Naval reactor fuel assemblies.

That's where things went awry. Although nominally zirconium-clad, the Navy's reactor fuel assemblies were overplated (alloyed?) with a secret element that no one bothered to tell the Chem Plant's fuel dissolution experts/designers about (and they didn't ask). Anyway, that little technical detail served to render dissolution of those fuel assemblies far more difficult than had been assumed, which rendered the entire facility's productivity so poor that the Navy finally decided to cut its losses[208] and just continue to "temporarily" store its spent fuel at the Site. However, the State of Idaho objected, demanding that the DOE develop a plan for managing both those fuel elements and all of the already-generated reprocessing wastes, both liquid and solid, plus various other waste materials scattered across the remainder of the INEL. A lawsuit to prevent any further receipt or storage of spent fuel until such a study had been completed accompanied that demand. In 1995 this conflict resulted in an agreement between DOE, the State of Idaho, and the U. S. Navy outlining the future of fuel storage and waste management at the INEL. DOE promised to "clean up" everything at the Chem. Plant including the roughly one million gallons of still-liquid reprocessing wastes within several of its then >40 year old tanks. In 1998, DOE identified the Chem Plant's Main Processing Building (CPP-601) for removal, along with its Fuel Storage Building (CPP-603), and several support structures. In 2005, DOE deemed most the remaining ICPP structures including its waste calcination facility (NWCF) obsolete and the deactivation, decommissioning, and demolition (DD&D) of its

[208] DOE/INEL had asked the Navy for another half billion dollars to replace its Fluorinel Dissolution Facility's almost-new dissolvers with "safer" ones but the end of the cold war had cut defense-type spending to the point that even the Navy's pockets weren't deep enough to keep that project going (it wasn't going to recycle that uranium back to its own reactors anyway).

buildings and other structures began, including everything associated with fuel reprocessing.

This left the folks who had worked there without much to do other than watch DOE's subcontractors bulldoze everything into pits and pour gravel and/or concrete over it. The exceptions were the folks given an opportunity to try to get INL's infamous "steam reformer" (IWTU) up and running. As of this date (March 2019) they've spent 14 years and about a billion dollars trying and haven't succeeded yet.

Anyway, the INL's Chem Plant was built to do what virtually every other fuel processing plants does; i.e.,

1. Receive and store spent fuel until it's "cooled off" enough to work with – typically at least one and more often, ~five years
2. Dissolve that fuel in a way that generates an acidic aqueous solution containing several molar nitrate ion (unlike most civilian fuel reprocessing facilities, the Chem Plant dissolved cladding along with the fuel "meat" thereby generating solutions containing whatever the cladding consisted of along with the uranium, plutonium, and fission products).
3. Extract the uranium and plutonium from that solution via countercurrent liquid-liquid extraction with ~30% tri-butyl phosphate in a kerosene-like organic solvent. (uranium/plutonium are extracted as zero-charge nitrate-based, "ion pairs", e.g. $UO_2^{+2}(NO_3^-)_2$. This is a key step of the "PUREX" process)
4. Transfer the U/Pu in that solvent to a second aqueous phase by contacting it with water containing very little nitrate (e.g., 0.1 M HNO_3 this breaks up the ion pair rendering the U/Pu insoluble in the organic phase).
5. Boil off most of the water in that solution to concentrate it, add aluminum nitrate to bring the nitrate concentration up again, add some hydroxylamine and/or ferrous sulfamate to reduce/convert quadrivalent Pu to unextractible trivalent Pu, and then extract the uranium from it, again as an ion pair, via countercurrent liquid-liquid extraction with/into "hexone" (methyl isobutyl ketone). This "2nd cycle" hexone/aluminum nitrate-based process is called the "REDOX process".
6. That hexone phase was first rinsed with more aluminum nitrate to remove contaminants and then contacted with dilute nitric acid to transfer the uranium back into another aqueous solution.

7. To generate an even purer uranium product, steps 5 and 6 were generally repeated (this was the Chem Plant's "third cycle" extraction)
8. Most of the water in that now very pure uranium nitrate solution was then boiled off after which it was then squirted into a small fluidized bed "denitrator" which converted it to a compact, easily handled, dry UO_3 powder ready for shipment off to a customer (usually a fuel fabrication facility).

The wastes generated by such reprocessing consisted of:

- First cycle raffinate (aka "high level waste"): the aqueous phase coming out of the first cycle PUREX extraction. It contains the majority of the fuel's fission products and cladding components (e.g., dissolved steel or aluminum or zirconium) along with whatever else was added to either enhance extraction or stabilize the solution (often aluminum and/or calcium nitrate salts).
- Second and third cycle raffinates: these solutions contain lesser levels of fission products, almost all of the plutonium (generally), plus lots of aluminum nitrate.
- Miscellaneous clean up wastes: solutions of sodium (or potassium) carbonate, permanganate, and/or sodium hydroxide were used for cleanup activities throughout the plant primarily because "caustic" readily dissolves the silicic acid gels that gummed up pipes and extraction equipment[209]. Once so dirtied, they usually ended up in the "sodium bearing waste" tanks along with other small volume miscellaneous waste streams. A good deal of the mercury utilized as a catalyst during the dissolution of aluminum clad fuel elements also ended up in them.

---

[209] The silicic acid originated from the "purified" ~20 ppm silica ground water utilized to make solutions. The water purification process utilized (dual bed ion exchange) didn't remove silica very well because it is largely present in ground water as non-ionic silicic acid. Currently, that silica comprises the bulk of the sludges underlying the liquid in INL's SBW tanks. If INL ever does get its steam reformer going, the current plan is to leave them there along with a hefty fraction of that waste's total radionuclides, esp. ^{137}Cs & Pu. (those tanks are are to be pumped until "suction is lost").

The PUREX process utilized by most of the world's fuel reprcocessing plants entails

- chopping spent fuel rods into short pieces
- rolling/crushing them to break up/remove their brittle uranium (mostly) oxide pellet contents
- dissolving that ceramic in strong nitric acid
- extracting actinides (mostly U+Pu) from that solution via counter current liquid-liquid extraction with TBP/kerosene
- back extracting them into dilute nitric acid
- separating U from the Pu by adding a reductant (typ. hydroxylamine or ferrous iron) and more strong nitric acid, and extracting again with more TBP/kerosene - the uranium is re-extracted but Pu^{+3} isn't.

Excellent discussions of just about every aspect of LWR fuel reprocessing can be found in the "best" (second edition) version of Benedict, Pigford, and Levy's monumental "Nuclear Chemical Engineering" textbook [229].

Future Reprocessing

Because there are so many different ways that a sustainable nuclear renaissance might be implemented, I'm just going to describe a few scenarios.

For instance, a way to convert spent LWR fuel to something compatible with a MCFR would be to chop the fuel tubes into short sections, roll/crush them to break up the uranium oxide pellets; powder them; add one mole of $ZrCl_4$ per mole of total actinide; disperse both in 2 moles molten NaCl/mole actinide; and then slowly add zirconium powder while stirring. Thermodynamic calculations indicate that this should produce a fuel salt containing trivalent fissile/fertile actinides in the right amount of NaCl for the mixture to exhibit a melting point <550°C along with solid filterable/removable ZrO_2. The same sorts of calculations indicate that another way to do it would be to disperse finely ground UO_2 fuel pellets in molten NaCl and bubble carbon tetrachloride through it. The UCl_4 in the resulting $PuCl_3/UCl_4/NaCl$ solution could then be reduced to UCl_3 by stirring in one third as much powdered uranium metal.

Converting the same spent LWR fuel to something capable of starting a MSFR or tube-in-shell thorium breeder could probably be best accomplished via the same Purex based process I've outlined above. However, a great deal of money has recently been spent demonstrating that it could be accomplished via electrometallurgical "pyroprocessing" [102]. That process invokes electrochemical reduction of such fuel's oxide-form actinides (uranium along with some TRU most of which is plutonium) to metals after which the uranium and plutonium are separated via another electrochemical dissolution/redeposition process utilizing different electrodes. In this writer's opinion that approach is unlikely to be the "best" way to recover any such potentially useful plutonium fissile (the bulk of any such uranium would be stored for reuse after the reactor had been started). Pyrometalurgical processing's chief virtues are: 1) it could be performed upon "hotter" (not so much cooled-off) spent fuel than could any process utilizing radiation-labile water/organic extractants; and 2), it doesn't do a sufficiently good job of separating plutonium from uranium (or anything else) to generate something that imaginary terrorists could easily make a bomb "pit" out of (it'd be too "hot" to work with or sneak pass inspectors).

It would be more reasonable to fluorinate such spent fuel to volatilize the uranium along with some of the FP products and then subject the residue to a PUREX-type separation to recover fissile plutonium. That could be done by boiling that residue in concentrated nitric acid to remove fluorine, diluting that solution with water to generate another one that TBP could extract the Pu from while leaving most of the FP behind.

However, once such fuel has been made and partially "burned" in some sort of MSFR, electrometallurgical pyroprocessing might become a

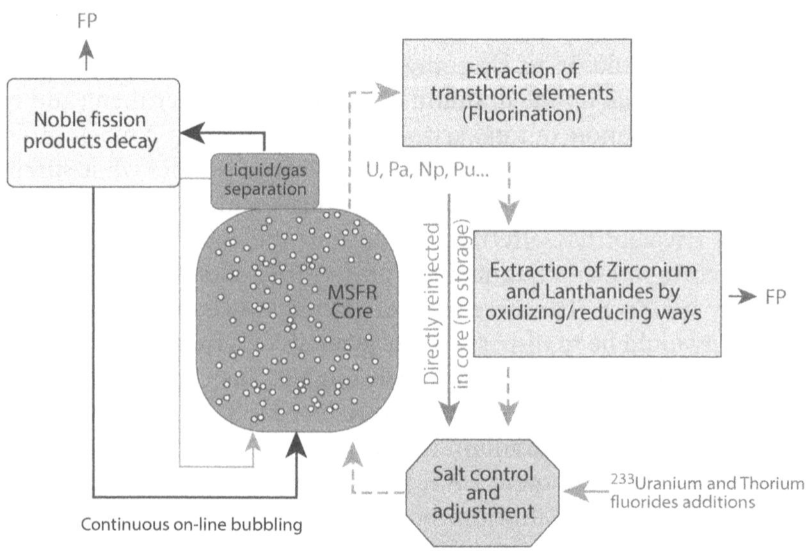

Figure 28 Generic MSFR fuel salt treatment scheme [215].

reasonable way of keeping it clean enough to remain useful[210]. See [230, 215] for descriptions of some of the salt cleanup schemes being evaluated for the MSFR.

The tube-in-shell reactor's fuel salt could be purified by first fluorinating the ^{233}U out of it (as UF_6 gas) and then distilling/collecting the FLiBe off the bulk of the FP under partial vacuum [231].

[210] Only 60-70% of the FP so formed would remain in any MSR's fuel salt because gasses (e.g. xenon, krypton, and tritium) and elemental metal "scum" formed from FP "noble" at its buffered/controlled redox state (e.g., Ag, Ru, Pd, Pt, Tc, Rh, Re, Sb, etc.) would be continuously sparged/skimmed out of it. Riley *et al.* have recently reviewed molten salt reactor waste and effluent management strategies (Riley 2019).

Appendix II. MSFR Isobreeder Fuel Salt Reprocessing

EVOL's neutronics modeling indicates that the 1.5 GWe MSFR could run at steady state (CR just over 1.0) with just 6 liters per day of fuel salt reprocessing (cleanup)[211].

The principle of "additive fractional molar volumes" (see APPENDIX IV) indicates that a fuel salt containing roughly 77.5 mole%, LiF and 22.5% of a mixture of $(HM)F_3$ and $(HM)F_4$.(HM = sum of actinide (heavy metal) nuclides), would have a density of ~4.28 g/cc and contain 47.9 moles/liter lithium plus HN fluoride salts.

Consequently, 6 liters of it would contain ~14 (0.93* 6*0.225*232* 47.9/1000) kg of thorium[212] and 1.59 kg (0.775*7*47.9/1000) kg of ^{7}Li.

Reprocessing (fuel salt cleanup) would involve sparging the fissile ^{233}U out of the spent salt (as UF_6, another gas) with elemental fluorine (or possibly NF_3) which would oxidize its uranium to gaseous UF_6. That uranium would then be returned along with ~14 kg of fresh thorium and 1.59 kg of fresh ^{7}Li (both as their fluoride salts) back to the reactor's fuel salt stream by reducing it back to UF_4 with hydrogen gas.

The spent fuel salt from which its fissile is so recovered/recycled could simply be discarded – at least temporarily – because there's nothing in it that's really worth the cost of recovery. If it is subsequently deemed worthwhile to recover/recycle its ^{7}Li, that could be accomplished by dissolving that salt in water and then adding ammonia – most of the FP and thorium would precipitate out as insoluble oxides/hydroxides leaving the ^{7}Li in solution as the much more water-soluble lithium fluoride. Boiling

---

211 The "reprocessing" referred to herein deals with "salt seeking" fission products, primarily alkalies, alkaline earths, and rare earths, not to what's continuously done to remove intrinsically insoluble fission products - a total of ~1 kg/day/GWe mixture of gasses and noble metal sludge – from the fuel.

212 At steady state about 93% of the HM in this system's core would be thorium; most of the rest would be quadrivalent uranium.

that solution to dryness would recover the ^{7}Li (as ^{7}LiF) which could then be refluorinated with elemental fluorine to remove any miscellaneous oxides and regenerate « fresh » ^{7}LiF. However, since AVLIS isolation of ^{7}Li from natural lithium shouldn't cost more than about $200 per kilogram [146], both ^{7}Li and thorium recovery/recycle could be put off until most of that salt's fission products have decayed away.

Appendix III. More Opinions about TERRAPOWER's Reactor Concepts

The total amount of natural uranium (NU) that has been mined to fuel the USA's fleet of commercial LWRs generating almost 20% of the U.S. electricity comes to approximately 700 thousand tons. Of those, about 70,000 tons ended up as spent fuel – the enriched uranium fed to the LWRs and then discharged after a few percent of it –mostly ^{235}U – had fissioned. More than 600,000 tons of that NU ended up as depleted uranium (DU) "waste". Additional DU was generated by military programs. In principle liquid metal (cooled) fast breeder reactors (LMFBRs) could burn nearly 100% of such "waste". However, that cannot be achieved in a single irradiation campaign because neutron-induced cladding damage constrains the burnup level achievable with their solid-type fuels to 15% to 20% FIMA (Fissions per Initial heavy Metal Atom), depending upon how fast the core's neutron spectrum is. Consequently, high uranium utilization requires multiple fuel recycling. Traditionally, that includes removal fuel cladding, separation/removal of most of the fission products from the fuel "meat", addition of depleted or NU uranium make up fuel, fabrication of new fuel elements and finally reloading them back into the reactor core for another irradiation cycle. Although technically feasible, there is a significant objection to fuel reprocessing in Western countries due to economic and proliferation concerns. Fast breeder reactors could, in principle, also operate without fuel recycling; that is, using a once-through fuel cycle as do all of the LWRs presently operating in the USA. Although a discharge burnup of 15% to 20% FIMA is 3 to 4 times higher than that of contemporary LWRs, a LMFBR's natural uranium consumption is not significantly different from that of a once-through LWR per cycle because the degree of uranium enrichment required to fuel a fast reactor is more than twice that required to fuel a LWR. Nevertheless, it may be possible to realize a significant increase in the uranium utilization by "reconditioning" fuel that's reached its radiation damage-limited degree of burnup. The functions of such re-conditioning are to remove a fraction of the fission products, primarily the gaseous ones, and replace the fuel clad prior to fuel re-use in the reactor. The objective of recent fuel reconditioning research is to overcome material performance limits in a way that cannot extract pure plutonium and is, hopefully, not as expensive as conventional fuel reprocessing. After reconditioning the re-fabricated fuel is loaded back into the core for additional burnup and thereby increasing uranium utilization [232].

According to Dr. Greenspan, Terrapower's initial Traveling Wave (TRW) Liquid metal fast breed and burn reactor (LMFB&BR) and its "Stationary Wave" LMFB&BR successor would operate on a once-through fuel cycle burning most of the Pu and minor actinides bred therein without separating them from the fuel. It's assumed that fuel cladding will last much longer than what's ever been achieved before after which it will be "reconditioned" in a way that doesn't isolate plutonium in a sufficiently pure form to be weapons useful.

To me it seems that that is just semantics: someone's decided to call the "reprocessing" of fuel that's reached its radiation damage and/or FP buildup limits, "reconditioning" instead.

That's too "lawyerly" as far as I'm concerned - rather like substituting "LFTR" for "MSBR" or "steam reforming" for "calcination". Employing slippery terms/definitions and generating deliberately vague documentation is how DOE's waste management experts managed to convince its Idaho stakeholders that steam reforming would be the best/cheapest/safest etc. way to convert INL's remaining liquid reprocessing waste to a competent radioactive waste form (see APPENDICES XII & IV).

I don't see anything in TERRAPOWER's MCFR patent applications that I consider "patentable" with the possible exception of their displacement-type "control rod". Another thing it/they seem to be claiming is the use of a mixed multivalent fuel salt (roughly equal amounts of tri- and quadrivalent uranium) in order to rationalize some of their apparently assumed very low salt melting points[213]. That'd be swell if thermodynamic calculations didn't suggest that any of the periodic table's metallic elements (molybdenum, nickel, platinum, palladium, etc.) probably would be oxidized by it, but unfortunately that's not the case[214]. There's a reason why almost everyone

---

[213] It seems that a better way of arriving at a sufficiently low melting salt mixture would be to substitute $MgCl_2$ for some of the NaCl. Beneš & Konings suggest that a 0.607 NaCl, 0.089 $MgCl_2$, & 0.304 UCl_3 (mole fractions) would melt at just 448°C (Beneš 2008).

[214] It may be possible to prevent such corrosion by deliberately maintaining an equilibrium level concentration of the relevant corrosion product(s) (e.g, Fe^{+2}) in solution. This is just one of the possibilities that only real-world experimentation can properly evaluate.

who's taken a serious look at the MCFR concept in the past has assumed that essentially all of its uranium must be trivalent resulting in a much more reducing – less corrosive – fuel salt. A mixed multivalent fuel salt work if someone comes up with a super ceramic material to make or line the reactor with, but I haven't heard about it yet.

Again I feel that a breed and burn MCFR is apt to be superior to any sort of solid-fueled B&B concept for several reasons. However, it's going to take more than what I've seen so far to convince me that TERRAPOWER has come up with the "best" concept.

Appendix IV. Example Additive Molar Volume Calculation

Let's determine how much plutonium would be in one liter of a room temperature salt mix containing 5 mole %$^{239}PuCl_3$, 28 Mole% ^{238} UCl_3 and 67 mole % NaCl.

salt	g/mole	g/cc	ccc/mole	
$Pu^{37}CL_3$	350	5.71	61.3 (350/5.71)	
$U^{37}Cl_3$	349	5.5	63.5	
$Na^{37}Cl$	60	2.17	27.6	
cc/mole salt = 0.05*63.5+0.28*61.3+0.67*27.6=			38.8	cc/mole
g/cc salt =0.05*5.71+0.28*5.5+0.67*2.17=			3.28	g/cc
one liter contains 1000 cc/38.86 cc/mole or			25.7	ΣSalts/liter
weight Pu per liter = 0.05*25.7*239 =			307.1	g/liter
weight Σ HM/l = 307.1+0.28*25.7*238 =			2020	g /liter

Due to thermal expansion both of these concentration figures would be about 15% lower at the reactor's ~700°C normal operating temperature.

Appendix V. QBasic Startup Fissile Program

```
REM « Assumes 400 LWRs for first 5 years, then 15 year build-out with
   initial startup fissile »
REM « After that all NU goes to building new 1 GWe breeders »
REM « breeder CR can be anything from 1 to 1.3 ; initial startup fissile
   (typ. 1900 t, max 2380), NUM = years run time (typ 81) & NU t/a after
   20 years to NUM »
INPUT « breeder CR is « , CR
INPUT « tonnes available startup fissile is «, START
INPUT « numbers of years run is », NUM
INPUT « Tonnes fissile per start is « , TON
RTOT=400 :REM « initial 400 1 GWe LWRs, etc.»
FOR X=1 to5 : NU=NU+64000 : REM « 64000 is tons NU /a required by
   400 state-of-the- art LWRs »

NEXT
FOR Y=6 to20 : NU=NU+64000 : RTOT=RTOT+START/TON/(20-6)

NEXT
INPUT « Tonnes NU mined per year for starting more breeders is »,
   MINEDNU
FOR Z=21 to NUM
RTOT=RTOT+MINEDNU*.007/TON+(RTOT-400)*0.78*(CR-1)/TON  :
   REM « 0.78 is tonnes fissile burned/GWe-year »
NU=NU+MINEDNU

NEXT
PRINT « Total NU after «, NUM , « years is « , NU
PRINT « Number of one GWe reactors then is «, RTOT
```

Appendix VI. A More Realistic Tube-In-Shell Thorium Breeder Reactor Startup Scenario

Since the world contains relatively little ^{233}U to start reactors with, it would have to be bred in reactors started with either ^{235}U or ^{239}Pu. Here's a concrete cxample:

Case 1 of Alexander *et al.*'s report [116] assumes the same reactor described in my text, running with ^{235}U instead of ^{233}U. Its fuel salt contains 33.8E+19 fissile atoms/cc and its CR is 0.5448 (it's not a breeder when run with ^{235}U but does make some ^{233}U). So, if we assume the same-sized external plumbing ($8m^3$) as my text's example's, it'd require 1.746 tonnes of ^{235}U to start and would generate 54.48% as much new ^{233}U in its blanket (170 kg) required to start another same-sized breeder. The best way to see what this means is to write up a little basic program like that in APPENDIX V (see below)

INPUT "input the ^{233}U fueled breeder reactor's CR", CR (any number between 1.00 and 1.19)
INPUT "run time in years", years
Orig = 64000*0.0071/1.746: REM orig is the number of reactors that could be fed/started with the ^{235}U in 64,000 tonnes of NU (about 260 per year)
Increm = 0.5448*orig: REM Increm is the number of new breeders that could be started with the ^{233}U in the "orig" reactors (about 142 per year)
FOR x=1 to years
N=N+increm
N=N*CR (adjusts the number of ^{233}U reactors upwards if CR>1)

NEXT
PRINT "total number 1 GWe reactors in "years, "is" , N+increm

e.g., for a 50 year run time and a CR=1, we end up with 7349 reactors utilizing the ^{235}U in the same ~64,000 tonnes per year of natural uranium currently consumed by today's ~400 LWRs. If their CR were 1.05, we'd have 31,427 reactors after 50 years. If it were 1.19, we'd have over 30,000 of them in just 34 years.

Over time, breeding is like the "magic" of compound interest in that it can make you "rich".

I've assumed high-enriched uranium (HEU ≥90% ^{235}U) starting fissile because low enriched uranium (LEU) within the core would destroy its neutronics and thereby render the system unsustainable. Enriching natural uranium from a LWR's 5%^{235}U LEU to HEU, would raise enrichment costs by about 23% ([229], p, 671).

Here's another way to look at it: If it takes 0.312 tonne ^{233}U to bring the breeder version of this thing up to criticality and a ^{235}U fueled version of it has a CR of 0.5448, it should take 0.5776 (0.312*235/233/0.5448) tonnes of ^{235}U to start one breeder.

For 30000 such breeders that'd take the ^{235}U in 2.443E+6 (0.5776*30000/0.0071) tonnes of NU.

That's pretty close to the same amount of NU (400*160* 34 =2.18E+6) predicted by my BASIC program assuming a CR of 1.19.

It could also be started up with "excess" bomb grade Pu though doing so would somewhat complicate "reprocessing" because, unlike uranium, it can't be easily separated from either FLiBe or 25%ThF_4/75%LiF by fluorination.

Appendix VII. Letter Sent to INEEL's Director circa 2001 (After "Separations" & Before "Steam Reforming" was the Site's "Preferred Alternative")

"Ethics & EM – a view from the trenches"

Work's been pretty slow out at TRA during the last month. The reactor is down again, another ISM audit is pending (audits always generate paralysis), the ventilation system in my lab hasn't worked for three weeks (INEEL doesn't seem to be able to maintain blowers anymore), & my usual customers (I'm a chemist) are putting off doing things like submitting samples or doing experiments until the new fiscal year's funding situation settles down & the auditors go home. Consequently, I've had lots of time for "training" – most recently, "Ethics". For some reason, being reminded about "waste, fraud, & abuse" always gets me pondering about INEEL's EM programs. Since such thoughts aren't programmatic & I've already burned up too much of my boss's overhead, I've decided to take the day off (PL) to write you this note.

About two months ago I was invited to spend a week in town with about 30 other guys from INEEL and elsewhere (the outsiders included many of the DOE Complex's glass experts) to try to come up with a sure-fire way to implement "direct vitrification" of INEEL's sodium-bearing waste (SBW). We were told that "direct vit" would be the "preferred alternative" in the upcoming "final" INEEL HLW EIS & were presented with a "90% draft"

INEEL engineering document that goes into excruciating[215] detail about one way to do it (slurry raw SBW with glass forming agents & sugar in tanks, "characterize" those mixtures, & then feed them directly onto the cold cap of a SRP-type melter @~75 gallons per hour.)

We were also told that we were permitted to think about how the system might be used to treat existing calcines as long as those thoughts wouldn't impact on how SBW could be treated ASAP. We were divided up into subgroups that were to concentrate upon one aspect of the proposal.

I was probably assigned to the "waste pretreatment" group because I'd just written/distributed a note about the results of the steam-reforming experiment recently run (by SAIC) to verify contentions made in a STUDSVIK proposal that had been sent to me[216] three years ago.[217]

Just before we went off to our cubicle, I asked the visitors whether or not calcination (a.k.a. "drying", "denitration", "incineration", "steam reforming", etc....) is apt to make the job of converting dilute (solids-wise) nitrate-based liquid into glass easier. They were unanimous, "yes". Consequently, my subgroup (all from INEEL) proceeded to sketch out a scenario that would feed a slurry of SBW, sugar, glass forming agents[218] into a STUDSVIK-type steam reformer & blow the product ash/calcine directly into the melter[219].

When we presented our brainchild to the whole group, it was considered "radical" because what would be going into the melter would have been "characterized" one step before where the folks running the melters at SRS & WVDP do it.

---

[215] INEEL has a hoary tradition of "paper engineering" projects before doing the R&D needed to test key assumptions (that's how we built a $350 M fuel dissolution facility (FDP) that never achieved better than one-third of its promised throughput). The document I'm alluding to here specifies the size of every room of the proposed facility to the nearest inch & estimates costs to 5 - 7 significant figures. Don't you think it'd be better to determine (experimentally) if we really need (for example) five feed mix tanks before we go into this degree of detail on the paperwork?

[216] I'd previously written a note to STUDSVIK's Brad Mason (he'd recently managed a successful fluidized-bed SBW sugar-calcination project for Hanford) pointing out that INEEL would probably not honor its promise to calcine SBW unless an "outside" organization with real credibility & expertise stepped in with "an offer we couldn't refuse" (or ignore).

[217] Which I first passed on to INEEL management & then, when nobody paid attention to it, to INEEL Over Sight's, Kathy Trever.

[218] not "frit"- too difficult to keep in suspension – instead, easily suspended/pumped stuff like diatomaceous earth, jeweler's rouge (Fe_2O_3), & lithium tetraborate.

[219] Basically all we had to assume is that what goes into the calciner comes out the other end - minus "volatiles" of course. The British & French screwed up enough courage to make that assumption over a decade ago.

This brings me to one of the reasons for this note – the mind-numbing "conservatism"[220] evinced by most DOE Complex decision-makers & technical experts. By this I mean unquestioning acceptance of silly (& often blatantly self-serving) assumptions[221] that render accomplishing anything unnecessarily difficult, grossly expensive, and often impossible. For instance, in this case, the group was ignoring the fact that Cogema & BNFL have been successfully making genuinely "hot" (100's of times more radioactive than DOE's) high-level glasses for over a decade. European melters are fed by close-coupled rotary kiln calciners that simultaneously mix waste with glass forming additives and remove "volatiles" (water, free acid, nitrate, etc.). The reasons why the Brits & French do it this way include; 1) the removal of volatiles improves the productivity of their melters, 2) it simplifies offgas cleanup, & 3) it's intrinsically easier to accurately meter stuff going into a direct coupled calciner/melter system than it is to make calcine separately, "temporarily" store it, retrieve it, mix it with frit, "characterize" the mixture, and then meter it into a melter. In other words they do it that way because that's what both common sense & economics dictate.

Why would doing it that way here too be "radical?"

In the US, the rationale for saying that the stuff going into our melters has to be immediately "characterized" beforehand is that we assume that it's impossible to adequately premix our wastes. At INEEL this boils down to accepting the notion that low viscosity solutions containing very little undissolved solids situated in adjacent tanks couldn't be homogenized by simply pumping them back and forth for a few months (years?) – a patently ridiculous assertion.

If one has the temerity[222] to ask **why** "we can't" mix those liquids (I did), you're told that the **only** way to pump them back and forth would be to do so with steam eductors ("jets") & since this would dilute waste with condensed steam & the tanks are already "full", we "can't" do it.

---

220 For instance, the melter subgroup's technical guru (an outsider) volunteered that an existing high temperature pilot plant melter at his home base could be derated to approximate the characteristics of DWPF for R&D on our problem - do we really want another DWPF?

221 For example, for the last ten years, INEEL's HLW gurus justified their program with the contention that a several cubic mile mountain (YM) would be "too small" to contain 4200 m^3 (0.000001 $mile^3$) of calcine unless it's first been "separated".

222 Asking "why" in this business automatically makes you "not a team player" – which, in turn, negatively impacts your career.

Do you have any idea of how many different ways of pumping liquids have been developed? Couldn't a "lead lab" adopt one that **would** allow it to mix its waste?

Some of the other opinions recently expressed by INEEL's in-house technical/regulatory/safety experts include:

1) "We can't" efficiently calcine SBW in NWCF[223]
2) "We can't" recover/recycle NOx from NWCF offgas
3) "We can't" recover mercury from NWCF offgas

There's plenty of evidence that those assertions are all dead wrong. More important, DOE-ID's acceptance of "we can'ts" from its contractor's in-house experts has repeatedly resulted in its failure to accomplish things it's promised to do; e.g., 5 years ago, we solemnly promised our then-new Governor (of Idaho) that we'd calcine SBW.

A "lead lab" should keep its promises – if it doesn't, it'll lose credibility, then missions, &, finally, jobs[224].

Another "we can't" has to do with INEEL's refusal to earmark R&D funds for any HLW management scenario inconsistent with the notions that, a) physical size is the most important characteristic of high-level waste, and, b) glass-making is the "best" way to make any so labeled wastes "road ready"[225]. Consequently, when DOE-ID finally realized (it took eight years!) that "separations" didn't make much sense, the only alternative that

---

[223] The key to efficiently calcining SBW in NWCF (R.I.P.) is precisely what STUDSVIK is now proposing to do (& which the British/French already do) – add sugar. One of the reasons INEEL recently lost (abandoned) its ability to calcine waste is that its decision-makers refused to implement sugar calcination (their excuse was "safety issues"). Sugar would have also greatly reduced the amount of NO_x dumped into the atmosphere.

[224] For example, the shutdown of WERF & NWCF not only affected workers at those facilities but also people who provided services for them elsewhere; e.g., I used to do a good deal of WERF's analytical work at TRA. Since we're all still on the INEEL payroll, it's unlikely that abandoning those facilities/technologies/missions gave much relief to US taxpayers.

[225] The rationale for this "we can't" is usually, "*there's not enough money to look at more that one thing at a time*". The fact is that the resources needed to do such research here at INEEL has already been paid for - the equipment, the lab, & my time will cost US taxpayers the same whether they're put to constructive use or wasted. INEEL's bean-counting approach to time/personnel/project management prevents its senior scientific professionals from being more productive because research that's not been "stovepiped" all the way down from HQ can be construed as "time card fraud". It's much safer for individuals to occupy themselves with "training" during slack times.

INEEL had generated enough "official paperwork"[226] to fall back upon on was "direct vit".

DOE-Hanford's latest public relations disaster[227] ought to be a lesson for INEEL. Hanford can't afford DOE-style vitrification & neither can we.

INEEL could become a genuine "lead EM lab" if its leadership permitted it to. One way to make it happen would be to take the blinders off your technical people. There's no valid reason why the research I've done since 1995 (due in large part to the good offices of the Inspector General & an extremely patient wife) couldn't have been officially sanctioned/funded[228]. If it had, a lot more could have been accomplished, the tenor of the reports/publications generated would have been different, and INEEL would have something other than "direct vit" on its plate now. I'd also have a legitimate charge number to fall back on during times like this.

P. S. I've written two more "hobby" papers[229] about how & why INEEL could implement a more productive approach to rendering reprocessing waste road-ready. If you're interested, I'll be happy to send you copies.

---

226 "Official" is the key word here. The bibliography of the recent "Draft INEEL HLW EIS" includes only officially sanctioned/produced documentation, mostly DOE Reports. Relevant information from other sources (e.g., the National Academy of Science) apparently doesn't count. Could the reason for this be that much of the unofficial documentation doesn't support the paradigm adopted by the EIS's authors?

227 i.e., BNFL's first being publicly humiliated & then "fired" for "committing truth" about the cost of vitrification in the DOE Complex.

228 Here's a thought. How about giving your PhD's, let's say, 60 hours per month of "Professional Development" time to devote to R&D *they* consider to be consistent with DOE's professed missions/goals. The number/proportion of "Professional Development" hours should go up every time we abandon another "real" mission/facility, there should be no "carry over" limit on them, & we (the PhD's) ought to be able to trade 'em back & fourth to pay for in-house expertise/services.

229 Both were presented at this year's ***American Ceramics Society*** radwaste symposium.

Appendix VIII. Letter Sent to Two of DOE's Inspector General's Lawyers Just after My Job Had Been Downsized for the Last Time

(there was no response or any other action taken by them or anyone else in DOE's/INL's chain of command – I retired one year later. Most of the names have been X'd out to protect the guilty)

From: Darryl D Siemer [mailto:SIEMDD@inel.gov]
Sent: Thursday, April 07, 2005 10:32 AM
To: XXXXXXXX X XXXX; XXXXX X XXXXXX (the lawyers)
Cc: (*everyone in my chain of command*)
Subject: more HLW-related waste, fraud, & abuse

The latest YM imbroglio has heightened public awareness of the same issue that I've been discussing with you recently; i.e., can folks trust DOE's "science"? Here's some more background on INL's own little EM-science scandal.

At the beginning of this year one of our calcination experts gave me a copy of the fluidized bed steam reforming (FBSR) "book"[1] that THOR's chief technical expert had cobbled together a few months earlier (March '04) for a California-based consultant. The cover note prepared & sent along with that "book" indicated that it was being cc'd to DOE-ID's HLW/SBW treatment czar, XXXX XXXX (& only to him).

THOR's "book" made for fascinating reading (to me anyway) because it was the first time that a "proprietary" document describing what had really happened during HAZEN Research Inc.'s 2001 "FBSR mineralization" tests had been released; i.e., how its reactor was really configured, the fact that most of its product (also) consisted of a bulky, flour-like, dust (fines) - not "grapenuts", and that it (also) quickly rocked-up when they tried to produce a sodium silicate-type product. It is also fascinating (to me anyway) because it contains the first document I'd seen[2] which discloses information consistent with the "controversial" papers I'd written about INEEL's initial attempt at applying THOR's process to its sodium-bearing waste (SBW)[3,4].

Had I been permitted to see this book when Mr. XXXX (*DOE's local HLW/SBW treatment czar*) first passed it on to BBWI's fluidized bed experts, the job of defending the paper I'd just then written up for the 2004 American Ceramics Society's radwaste symposium would have been much easier. The reason for this is that it seems that no one else (including the reviewers picked for my paper) had apparently seen an unfiltered version of HAZEN's report either - as far as we/they knew, its reactor had produced only a dense, "grapenuts-like", product which official DOE-lab reports had characterized as "better than glass".

Consequently, my paper which suggested that ...

- Mineralization-style FBSR is not an intrinsically "easy" (cheap) process to run
- It invariably produced more dust than "grapenuts"
- It could not turn SBW into sodium silicate, and ...
- Its product is not as "durable" (leach resistant) as previous reports had indicated

... required lots of "defending". (I did, however, eventually prevail – the paper was published[4].)

The folks who had volunteered to run ACERSOC's radwaste symposium last year (& this year's too) just happen to work at the same DOE lab (SRS) which had provided THOR with the official reports used to "sell" its process[5]. By another coincidence, that same lab had recently become DOE's "lead lab" in managing its own steam reforming R&D program - this in spite of the that fact that SRS possesses neither the expertise nor the equipment required to properly evaluate a fluidized bed-based technology[6].

Curious.

Even curiouser, it seems that as soon as I began to point out some of the inconsistencies between what SRS had published about THOR's process/product & what I had observed during the local test, INEEL's decision makers decided that they could no longer afford to pay me to provide real-time process (chemical) support during the subsequent FBSR tests to be run for their new customer (SRS). They couldn't afford to pay me to continue to do post-run product characterization work either. (They subsequently also decided that I wouldn't be retained by the lab-side of the new/improved INL either - that's why I'm now a lab-less "cubicle dweller" here at TSB.)

Anyway, last Spring I volunteered to do "free" post-run characterization of the products of INEEL's CY 2004 FBSR demos & consequently was given samples of everything produced during the rest of those tests. I'm ATTACHING the texts of the notes I sent along with my free analyses/characterizations. Of course, since my results weren't "official", they were not included in the formal INEEL reports produced after each demo[7]. Another reason why my reports/conclusions didn't show up in INEEL's official reports was that its customer had dictated that all such characterization work be performed at SRS - not here where the stuff was being made. (You'll note that I repeatedly chided our team for not duplicating the conditions outlined in THOR's description of HAZEN's test[2]; i.e., implement 100% recycle of fines back to the reactor. One of the things revealed in the before-mentioned "book" is that we had indeed duplicated HAZEN's experimental system (almost anyway). What we did differently was to report everything that happened when we tried to run it.)

Everything I've seen indicates that FBSR would be inferior to a properly-implemented vitrification system for INL's SBW.

- FBSR achieves little or no volume reduction (about 6 x less than vitrification)
- it produces more and dirtier (dustier) off gas
- the process itself is apt to be problematic a rocked-up "mineralized" reactor bed couldn't simply be dissolved-out with dilute nitric acid
- its product doesn't meet the durability standard that waste forms made from similar radwaste at other DOE sites must satisfy (e.g., Hanford's LLW)
- its product doesn't meet CFR 10- 61 disposal/transport criteria (i.e., 'no dust')
- its product doesn't meet the YM WAC
- and, of course, "stakeholders" on both ends of the waste pipeline prefer vitrification

These observations shouldn't be too surprising because an independent "Technical Review" had recommended that DOE not attempt to produce a "repository ready" waste form with THOR's process back in 2000.

Copies of SRS's FBSR product characterization reports were ATTACHED to the note that I sent to you last Thursday[9]. As that note indicates, it's apparent that some highly-placed individual (i.e., someone who can decide who receives DOE's EM R&D funding) decided to champion THOR's technology several years ago and that SRS decided to go after that money. It's also obvious that INEEL's decision makers agreed to keep the lid on what was going on[10].

It's unfortunate that DOE decided to waste so much time & money on this particular EM boondoggle. It's more than just "questionable" ethics - the fact that INEEL was not permitted to continue to seek more efficient ways of vitrifying its waste while DOE was encouraging its contractors to "study" FBSR (the status quo approach to vitrification leaves plenty of room for process improvement[11]) means that INEEL may not succeed in converting its "old" reprocessing waste into a stakeholder-acceptable waste form[12]. Since its unresolved "waste issues" continues to be one of the US nuclear power industry's greatest handicaps, such failure would undermine INL's chances of playing a leading role in reviving that industry (which, of course, would generate "new" reprocessing waste).

While it's understandable why individuals whose livelihood depends upon remaining in the good graces of our decision-makers might be reluctant to speak up about issues like these, that's not the situation with me - I'm old enough to retire & have already been "downsized" out of several jobs here at INL.

Where do I go from here? Idaho's Congressional delegation?

1. "Steam Reforming Technology for the Denitration and Immobilization of DOE Tank Wastes, Book 1", Brad Mason, dated 19 March 2004, (a compilation of ten documents)
2. THOR's 2003 "Tuscon Conference" paper (J. B. Mason *et. al.*, "Steam Reforming Technology for Denitration and Immobilization of DOE Tank Wastes", WM'03 Conference, Feb. 23-27, 2003, Tuscon, AZ.) is a typical example of prior descriptions of the "HAZEN test" in that many of its conclusions are inconsistent with the information in the actual report (doc. TR-SR01-1, Rev 0, "Hanford LAW Waste THORsm Steam Reforming Denitration and Sodium Conversion Demonstration").
3. D. D. Siemer, M. W. Grutzeck, and B. E. Scheetz, "Cementitious Solidification of Steam Reformed DOE Salt Wastes", ECI Alternative Nuclear Waste Forms, Girdwood, AL January 18-23, 2004.
4. D. D. Siemer, "Steam Reformation of Sodium Bearing Waste: Pros & Cons", Ceramics Tranactions, Vol. 168, pp. 81-90, 2004.

5. THOR's characterizations of FBSR calcine as "better than glass" are generally referenced to SRS's reports. The one referenced most often is C. M. Jantzen's, "Engineering Study of the Hanford Low Activity Waste (LAW) Steam Reforming Process," WSRC-TR-2002-00317, Westinghouse Savannah River Company, Aiken, South Carolina, July 12, 2002.
6. For example, early last year the materials science experts supported by SRS's steam reforming project wrote up a formal description of how they proposed to spend their research money that year. The centerpiece of their proposal was an multiple step scheme to make a "mineralized" steam reformer product with more or less standard laboratory equipment (calcination in a muffle furnace, "hydrothermal" conversion of that calcine, etc.) - not in a fluidized bed reactor. When I pointed out to SRS's "project lead" (Bill Holtzsheiter) that the experimental conditions proposed for that project wouldn't emulate what happens in a fluidized bed reactor (THOR's system), the plan was modified as described in the project's final report (WSRC-TR-2004-00560 Rev A, "Evaluation of Fluidized Bed Steam Reforming (FRSR) Technology for Sodium Bearing Wastes From Idaho and Hanford Using the Bench-Top Steam Reformer (BSR) (U)" December XI (sic), 2004.) The modified scheme involved dribbling a slurry of clay, charcoal, "catalyst", and SBW into a big crucible situated in a preheated muffle furnace. The crucible contained aluminum oxide "bed" granules which were initially fluidized with steam. When the feed mix was added to it, the alumina bed quickly agglomerated (fluidization stopped) which resulted in the formation of a "giant turd" waste form (my appellation) - not a mixture of "grape nuts" and fines.
7. For example: N. R. Soelberg, D. W. Marshall, S. O. Bates, and D. D. Taylor, "Phase-2 THOR Steam Reforming Tests for Sodium-Bearing Waste Treatment, INEEL/EXT-04-01493, Idaho National Engineering and Environmental Laboratory, Idaho Falls, Idaho, January 2004.
8. J. A. Gentilucci, J. E. Miller, R. L. Treat, and W. W. Schultz, "Technical Review of the Applicability of the Studsvik, Inc. THOR Process to INEEL SBW" Rev. I (a "Tanks Focus Area" doc. published Sept. 2000).

9. "SRS's Steam Reforming Reports - another critical issue", Lotus note to XXXXXX & XXXXX (the lawyers), 31Mar05 (4 ATTACHMENTS).
10. For instance, our leadership decided to not report that the stuff produced by INEEL's first SBW mineralization test didn't exhibit the anticipated degree of "durability" - instead that report (INEEL/EXT-04-01493) indicated that PCT results were "still being evaluated".
11. For instance, the SBW vitrification system outlined in DOE's ICP RFP "shared documents" website (INEEL/EXT-04-01692) assumes that we would have to build another huge DWPF-type melter. That's not true because glass-making technologies have become more efficient since SRS's melter was designed.
12. Why? That book's "proprietary" HAZEN test/report (TR-SR01-1) was actually paid for (& therefore "owned" by) WGI/Bechtel National Inc. I probably don't have to remind you that both of those contractors have "proprietary" interest in this Site too.

Appendix IX. Suggestions for Improving INL Reprocessing Waste Management

INL's Already Calcined Reprocessing Waste

The fuel reprocessing systems developed by the US federal government's cold war nuclear defense plants to recover uranium and/or plutonium generated a far greater volume/mass of radioactive waste per unit mass of fuel processed than do those utilized by modern commercial fuel reprocessing facilities. US reprocessing radwaste is less radioactive because: 1) its defense-type reactor fuel was generally not "burned up" to nearly the same degree as are commercial fuels, which means that less fission products were produced per unit mass of uranium processed; 2) the entire fuel rod was usually dissolved (cladding as well as the fuel "meat") which further served to dilute the waste's fission products; 3) little attention was paid to minimizing the amounts of solid-forming chemicals (e.g., sodium, aluminum, and calcium nitrates, boric acid, cadmium nitrate, etc.) added to facilitate processing. APPENDIX XI discusses radwaste radiolytic heat generation.

Calcination of this sort of waste was often considered[230] because: 1) weight-wise roughly 80% of it (e.g., nitrate, nitrite, organic chelating agents, water, etc.) is intrinsically volatile; 2) calcination would facilitate its storage because calcines generally occupy considerably less space (typically 15-20%) than do the liquids from which they are made; 3) a calcine would not be as readily dispersed as a liquid if the storage vessel were to be accidentally breached; 4) the relatively low temperatures required to calcine waste (~500°C) would minimize co-volatilization of chemical toxins

[230] However, NRTS/INEL/INEEL/INL was only DOE site that actually did calcine its wastes.

(e.g., Cd) and radionuclides (e.g., ^{99}Tc & ^{137}Cs); and 5) calcination would facilitate subsequent implementation of the technologies (e.g., vitrification) which could create a better ultimate disposal form.

Fig. 29 is a schematic of the Chem Plant's fluidized-bed calciner. Waste solution was sprayed into a heated, air-fluidized, bed of granular solids where it first dried upon the surfaces of those solids after which its nitrate salts decomposed to form oxides. To begin with, its "starting bed" consisted of limestone granules – at steady state, the bed's particles consisted almost entirely of the waste's ash forming components. At that steady-state, feed deposition, particle elutriation, attrition, and deliberate removal of larger particles via overflow resulted in a steady-state particle size distribution. Its bed temperature was maintained at about 500°C via in-bed combustion of kerosene atomized by oxygen through wall-mounted spray nozzles. The calcined product consisted of a mixture of flour-like fines and sand-like granules which was blown through pipes to binsets (Fig. 22) several hundred feet away where it remains today. The system's dusty off gas was "wet scrubbed", passed through silica gel asdorbant beds, and finally HEPA filtered before reaching the atmosphere via a ~300 ft high stack.

The primary weakness of the purely thermal calcination process adopted at NRTS/INEEL was that it could not efficiently deal with wastes in which the molar ratio of alkali metals (primarily sodium and potassium) to polyvalent metal (aluminum, zirconium, iron, calcium, etc.) nitrate salts

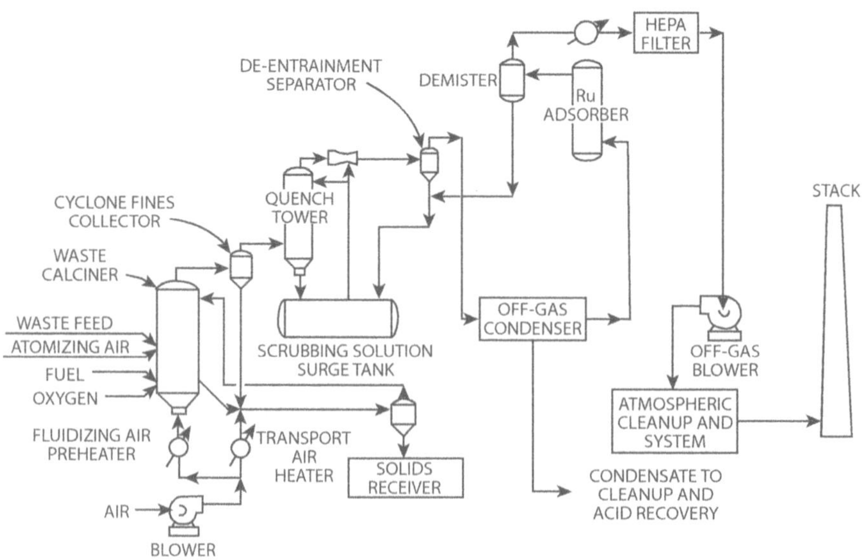

Figure 29 INL's New Waste Calcination Facility [233].

exceeded ~33%. The reason for this is that sodium nitrate does not thermally decompose at that system's operating temperature (~500°C) but instead melts to form a viscous liquid which agglomerated (defluidzed) the reactors bed. In 2000 (after DOEs separations waste treatment paradigm had finally run its course at INEEL), a combination of offgas-related stakeholder issues, INEEL's unwillingness to sugar calcine its remaining "sodium bearing" liquid waste" (SBW), plus the fact that its calciner was being heated in a way that apparently rendered it an unlicensed "incinerator" caused the site's decision makers to abandon calcination for "direct vitrification". Unfortunately, under two years later vitrification was abandoned too because DOE's topmost environmental management decision maker decided that it would cost too much to implement at INEEL.

The most sensible thing to eventually do (there's no great rush & no one should pretend that there is) with INL's calcined wastes would be to fill the reinforced concrete, vaults containing its stainless steel calcine « bins » with a blast furnace slag-based cementitious grout containing finely crushed phosphate rock and vermiculite. That combination of ingredients mixed with enough water to make a stiff but pumpable grout would set up to form a concrete capable of chemically immobilizing any fission product that might somehow find its way through the walls of those bins thousands of years from now.

If that's too simple, cheap, safe, and logical to be deemed acceptable, the same glass melter that could be used to make phosphate glass gems of INL's remaining liquid reprocessing waste (see the following subsection) could vitrify its calcines as well. Phosphate glasses are better-suited than borosilicate glasses for such things because they do a better job of retaining transuranic elements (mostly Pu) and anionic species (e.g., halides, $^{99}TcO_4^-$, and sulfate).

Better yet, we could slurry up those calcines with the raw SBW and phosphoric acid plus some powdered iron ore and melt the resulting « mud » to make phosphate glass « gems » of 100% of INL's reprocessing wastes. Doing so would reduce the total amount of glass required because those wastes' compositions chemically complement each other[231].

---

[231] A requirement for good quality, low/easy melting glass manufacture is that the relative amounts of glass network formers (silica and/or P_2O_5), alkali (sodium, potassium, cesium...) plus alkaline earth (calcium, etc) elements, and polyvalent metals (aluminum, ferric iron, zirconium, etc.) fall within certain ranges. Sodium bearing waste's non radioactive ash-forming components (Al^{+3}, Fe^{+3} & Na^+) would have to added to any phosphate glass made from just INL's calcines.

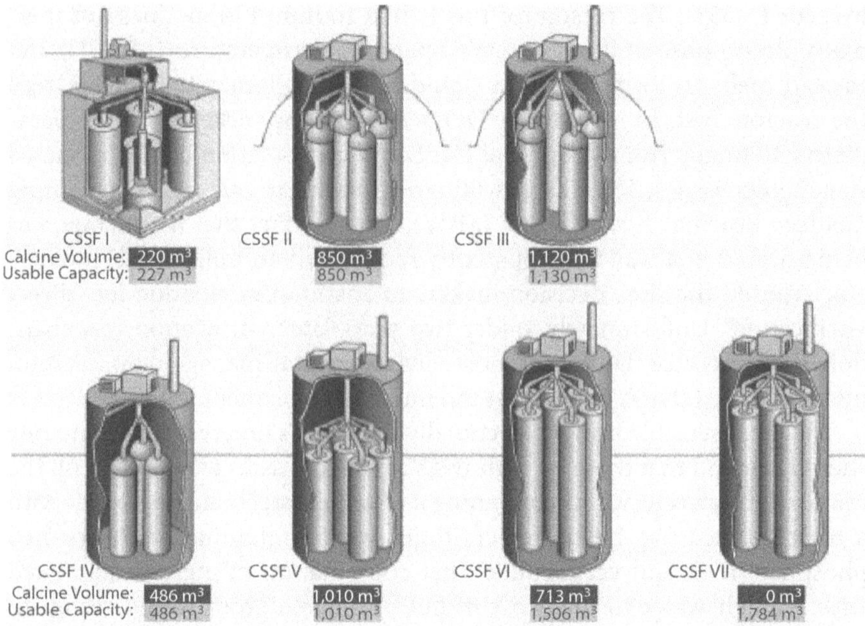

Figure 30 INL's calcine binsets ([228], CSSF VII is empty and big enough to contain anything that might be made of its sodium bearing waste (all of them are at least halfway underground).

As far as such glass's disposal is concerned, the most reasonable option would be to grout that gem-type aggregate back into the same binsets currently containing the untreated calcines. Failing that, we could ship them down to the NTS's Area 5 and bury them at the bottoms of a few more Greater Confinement Disposal boreholes (see APPENDIX X).

Of course, if that doesn't satisfy everyone and we don't mind waiting for a few more decades, I'm sure that a contractor will eventually promise DOE that it could slingshot them into the sun.

I recently mentioned two other disposal options for such glass in another QUORA answer. https://www.quora.com/Let-s-say-that-we-used-nuclear-energy-on-a-mass-scale-would-Mars-or-the-moon-be-good-resting-places-for-the-radio-active-waste-left-over (warning: some folks might be offended by those suggestions.)

Finally, since DOE has officially declared that INL's calcines will be hot isostatic pressed (HIPed) in one way or another before 31Dec2035, a more reasonable* way to go about doing it than is currently planned would be to mix them with sufficient dehydroxylated clay and/or class F flyash plus sodium silicate to make a stiff geopolyeric grout that could be pumped into the HIP cans. Such grout would set up to form a durable concrete which could either serve as the waste form as-is, or eventually be dried out and HIPed as promised [234].

***Although DOE has declared that INL's calcines will be HIPed in one way or another (see 75-FR-1 2010), it must first get permission to do so because HIPing does not satisfy the EPA's requirement for using the "Best Demonstrated Available Technology" (BDAT which currently means vitrification). HIPing is usually done by mixing dry powders (in this case, calcined waste with the additives required to form a durable product) and then transfering that mixture to the HIP container [235]. That's proven to be difficult to accomplish remotely without powder disaggregation, which causes the finished product to be much less durable than it could/should be. If everthing is done right (i.e., with the proper additives, a committment that DOE refuses to made), a HIPed product is apt to be more "durable" than a glass; i.e., last for even more eons after it's no longer radioactive enough to pose a threat to anyone.**

INL's Remaining Liquid Sodium Bearing Waste[232]

It's too late to do what should have happened over twenty years ago; i.e, sugar-calcine it [236, 237] and then store that product in INL's already built-and-paid-for binsets until someone decides for sure what's to become of its calcines. Since then its calciner (NWCF) was destroyed (decommissioned) over a decade ago and INL's subsequent steam reforming boodoggling hasn't accomplished much other than making work[233], its waste management decision makers should start over again from scratch. If I were in charge, I'd contact GTS Duratek and see if they might be willing to redo DOE's one and only « Vitrification and Privatisation Success » at its Idaho Site. SBW is compatible with that same glass formulation and the amount of it to be made is significantly less [154]. Since GTS Duratek *almost* broke even with its $13.9 million fixed-price bid circa 1995, it might be willing

[232] Almost 100% of INL's reprocessing wastes had some sodium along with its chemical-cousin, potassium, in them. INL's SBW had a higher proportion of such alkali-type metals (along with more plutonium and mercury) due to the fact that they had originated from "not first cycle" separation activities (in DOE-world, many things are defined in terms of what they're <u>not</u>).

[233] On 14Mar2019 I got another note (CNN 323337) from the subcontractor currently in charge of INL's steam reforming project (FLUOR), announcing that DOE is submitting another Class 3 permit modification request to Idaho's Dept. of Environmental Quality to make yet another set of 11 major and 1 minor change to that facility.

to do this project now for, let's say, $75 million which is surely under 10% of the tax dollars boondoggled away trying to get INL's steam reformer to work. The one thing that Duratek's CEO should insist upon though, is that INL's waste management gurus not be allowed to manage its project.

If in order to save face, DOE's decision makers insist upon using INL's billion dollar "reformer", I'd suggest running it under the same conditions that would have worked with NWCF:

- add jet-type mixers to the SBW tanks so that everything in them, not just supernates, are processed
- heat the reactor via on-bed combustion of kerosene rather than coal - don't bother with making/adding steam
- run it at about 500 rather than 700°C (doing so will mitigate « bark » formation)
- Mix sugar with the SBW squirted into the reactor calciner via a mixing « tee », not by premixing everything together in a big feed tank.
- Continuously blow its « carbonate product » (calcine) over into binset number 7.

If/when that reformer/calciner plugs up, do the same thing that we did with NWCF: shut it down, dissolve out the bed with nitric acid, return that solution to the SBW tanks, & then start up again. It should not take more than two years to complete the job.

Appendix X. Greater Confinement Disposal

The scientific basis of Dr Winograd's proposal was subsequently thoroughly investigated by SANDIA which renamed it "Greater Confinement Disposal". In 1984, DOE-NV acted upon this evidence and had a contractor (REECO) bury 500,000 Ci of ^{90}Sr and 200,000 Ci of ^{137}Cs in the lower 50-foot sections of two 120 foot-deep, 10 foot-diameter, holes augered into the NTS's Frenchman Flats (Area 5) (Dickmann 1989), [238] utilizing the same equipment used to emplace the "devices" which caused that region to appear pockmarked with smaller versions of Dr. Winograd's Sedan Crater repository site (see « GOOGLE EARTH »37.178N116.047W). The hotter of SANDIA's repository-holes contains roughly 3 kilowatts worth of fission product radioactivity – about one-thirty fifth of that in the INL's entire accumulation of calcined reprocessing waste. Approximately 1000 Ci of transuranic waste including 6 kg of weapons-grade plutonium along with hundreds of kilograms of uranium released by devices that had "fizzled" rather than explode, was buried in another set of four boreholes. The total amount of plutonium in all of INL's radwaste is about 35 kg - roughly 1% of the amount deposited both within and upon the NTS by DOD/DOE's weapons testing program.

This sort of cheap disposal scheme should eliminate the single greatest action-paralyzer/cost-driver in the USA's approach to radwaste disposal – the ridiculously high costs assumed for the waste's final burial plot.

SANDIA has since released several formal "performance assesments" (PAs) of DOE-NV's GCD [239].

They contain the following conclusions:

- No even-moderately-probable chemical/physical/mechanical mechanism other than direct deliberate human intrusion can cause that repository to fail within 10,000 years. [In this context, "fail" means to not meet the requirements of 40 CFR 191.]

- In spite of the buried fission product elements' (esp. ^{90}Sr & ^{137}Cs) far greater initial radioactivities and greater "leach-ability" as indicated by the leach tests that the developers of high-tech waste forms use to compare their wares at DOE-sponsored radwaste management conferences, those components do not dominate the hazard index of that repository - the much longer-lived « transuranic » (TRU) elements do. This means that short of an actual meltdown[234], any GCD disposal system (such as Winograd's) suitable for TRU wastes is "conservative" with respect to any fission products or chemically toxic stuff that might be mixed in with it. Most defense-type HLW is mildly "hot", somewhat chemically toxic (contains things like cadmium and mercury), and includes enough miscellaneous transuranic actinides to be deemed « teansuranic » (TRU waste) as well.
- The natural mechanism apt to cause the greatest degree of actual leakage of radiation from the GCD to the biosphere during the next 10,000 years involves the gaseous transport of trivial amounts of gases (radon and tritiated water vapor) upwards to the earth's surface - not the leaching of anything down to an aquifer situated more than six hundred feet below the burial zone. Geoscientific research has shown conclusively that there is not now nor has there been for at least 600,000 years (through several comings/goings of glacial ice sheets over most of the USA) sufficient water at the NTS to make the leaching of materials buried at appropriate depths an important dispersal mechanism.

Conclusion: In a competently-sited geological repository, the leach reistance of the waste form material itself is irrelevant.

- DOE-NV and SANDIA are so convinced of the validity of the GCD disposal concept that they have repeatedly and publicly advocated that it be seriously considered for the disposal of the DOE-complex's "orphan wastes"; i.e., wastes for which the existing set of assumptions/laws do not provide even "pretend" repositories like yesterday's Yucca Mountain [238].

234 If the burial crew is even remotely competent (heat generation is easy to meaure), a melt down is impossible.

How many boreholes would it take to make a competent repository for the INL's calcined reprocessing waste? If we assume that decision makers would not want to exceed the heat-loading of the "hottest" existing GCD borehole (three kilowatts) and also that the holes are the same size and depth as were the original ones, the number needed works out to be ~60 (see APPENDIX IX for a discussion of heat issues). The volume represented by the lower 50' sections of 60 individual, ten-foot-diameter, boreholes is 6660 m^3 – about the same as that of the concrete needed to "encapsulate" all existing ICPP waste.

What would such a repository cost US taxpayers? Sandia reported that each of the boreholes in its original GCD cost $20,000. If we apply a 5% inflation factor compounded for 38 years to sixty times $20,000, the cost of the proposed repository works out to $7.7 million dollars. That is approximately 0.4% of INL's current annual payroll cost.

Most of the reasons why an augered-shaft GCD repository would be "better" lie in the intrinsic nature of NTS alluvium. First, both because the soil in question is extremely dry and the little rain that does fall in « Area 5 » (about 10 cm/year) evaporates before it can reach a significant depth, convective transport of buried material is negligible. Diffusion, the only other natural process allowing readionuclide migration, is vanishingly slow in vadose-zone (dry) soils possessing high ion-exchange capacity. Second, plant populations are very sparse, not of the variety eaten by humans, and their roots don't go down 100 feet - so the food poisoning « what if » doesn't hold water. Third, because the NTS's alluvium accumulations cannot support open fractures ("cracks"), any future tectonic activity in that area would have minimal consequences. [This is not a trivial point because it means that no hypothesized "crack" could channel water from a hypothesized future flood directly through the burial zone to the underlying aquifer – floodwaterv water would have to diffuse downwards uniformly. Vulnerability to hypothetical "failures" caused by hypothetical cracks is a fundamental weakness of brittle, hard stratified rock, repository sites such as Yucca Mt, which is what made it so difficult (and terribly expensive) to "prove" that it could work for 10,000 years[235]. Fourth, the vertical-shaft design of a GCD repository minimizes the footprint of the disposal system which means that the probability of inadvertent human intrusion is minimized relative to mined repositories. Fifth, the geologic setting of the NTS's alluvial plains is such that potential future change, natural or man-made, would have little effect

---

[235] DOE's "trans scientific" YM modeling exercise never did (or could) succeed in "proving" that it couldn't fail any more than I/you could prove that a meteor couldn't possibly take out Airforce One next Tuesday.

on overall system performance. Sixth, the thick alluvial deposits in question not only happen to be situated in the driest part of the USA's driest desert, they have also been thoroughly contaminated by about 925 nuclear "events," over 120 of which were deliberately set off above ground. If there is any other place within the United States that's apt to be considered by the average US citizen/taxpayer/voter as a better place for a nuclear garbage dump, my studies have not yet revealed it. Seventh, it is simpler and therefore safer to emplace a drum (canister) of waste straight down a shaft and then backfill it than it would be to emplace the same object into one of the maze-like configurations usually envisioned for engineered geological repositories. Eighth, and perhaps most important, unlike the situation with Yucca Mountain, a GCD repository has already been implemented and already performance-assessed with real waste.

It would be both cheap and easy to improve upon DOE-NV's original GCD. For example, since it happens to be situated where the water table is relatively shallow by NTS standards (780 feet down), it cannot be fairly characterized as the most "conservative" site therein. [The portion of Yucca Flats that Winograd proposed for the nation's TRU repository has a water table 2½ times deeper]. The existing GCD was not sited at Yucca Flats because the same hydrogeological features that make it the best place to site a GCD also made it the most "defensible" place for DOD/DOE to do underground nuclear testing – the overriding mission of the NTS (nobody pulls more political weight at the NTS than does the US military). Secondly, SANDIA's performance modeling indicates that it would probably be better to bury wastes somewhat deeper than that existing GCD's 70'. Twenty years ago REECO engineers told me that the NTS possessed drilling rigs capable of drilling 12-foot diameter holes to depths exceeding 2000 feet (they were originally built to emplace big nuclear "devices" and test equipment). This means that waste forms could easily be buried at any depth considered optimal by Sandia's performance assessment modeling experts [probably below the 30-meter figure that the NRC considered to be the lower bound of "near surface"][236]. Next, since competent cementitious waste forms would be buried instead of raw waste, the disposal system should be credited for the well-documented retardation properties of the concrete matrix. [SANDIA allowed no "waste form credit" in its performance assessment of the existing GCD; i.e., modelers assumed that the waste is neither solidified nor "contained" within any engineered barrier]. Finally, it would be a very simple matter to pour a cementitious

[236] Most of the approximately 240,000 m^3 of radiactive waste buried in INL's "Radioactive Waste Management Complex" is covered with under one meter of soil.

grout around and over the so emplaced drums of waste before backfilling those holes. Doing so would meld the individual drums of concrete into huge monoliths having a much smaller surface:volume ratio to leach from and, more importantly, would) provide enough chemical buffering to stabilize the entire system for millions of years.

As a concession to Nevada's' political sensibilities, I should point out that if the cluster(s) of boreholes constituting such a new repository were to be drilled over existing bomb "chimney(s)", the concrete caps that would be put over them to direct rainwater off to the side would also direct that same rainwater away from the unremediated waste left by the original "events". It is also worth noting that waste disposal would provide a new and worthwhile mission for the hundreds of Nevadan site workers who are probably wondering how they are going to support their families now that the USA's cold war-inspired bomb testing is over (I hope) and DOE's Yucca Mountain study has ended. Finally, I should mention that no European nation has any such easy choices to make regarding the siting of their radwaste repositories. No EU nation is blessed with a large desertified region possessing the requisite hydrogeology and none have already contaminated large areas of their own soil with hundreds of nuclear blasts. The USA's decision makers ought to consider its Nevada Test Site (NTS) to be yet another of their country's unique natural assets and put it to productive use.

Appendix XI. How « hot » Are DOE's High Level Wastes?

The absorption of ionizing radiation by anything generates heat energy within it. Safe dissipation of such heat is a key criterion in the design of nuclear equipment, waste form materials, and waste repositories.

According to WHC-SO-WM-EV-102, REV 0, (« Single-Shell and Double-Shell Tank Waste Inventory »- Google it), in 1995, Hanford's roughly 54 million gallons of tanked reprocessing wastes had a total of about 3.3E+7 curies of ^{90}Sr and ^{137}Cs in them. Since there's 264 gallons per cubic meter, both isotopes have ~30 year half lives, and Hanford's wastes have decayed for another 24 years (i.e., is now 57% [0.5^(2019-1995)/30)] as hot as it was in 1995), this means that Hanford's HLW now contains an average of about 92 Ci/m^3 of ^{90}Sr and ^{137}Cs [0.57*3.E+7/54E+6/264]. Because the decay of both isotopes generates about one million electron volts (ev) (1.6E-13 Joule) and one Curie =3.7E+10 decays per second, that much activity translates to a heat generation rate of about one half watt/m^3. In light of the fact that a burrowing mouse generates more heat than that, it's not tough to understand why it would now be perfectly safe to grout glass gem « aggregate » made of such waste back into Hanford's tanks.

According to INEEL/EXT-98-00455, Rev 4 (Calcined Waste Storage at the Idaho Nuclear Technology and Engineering Center - Google it), >99% of its radioactivity is due to the decay of roughly 2000 Ci/m^3 each of ^{90}Sr and ^{137}Cs. That translates to a heat generation rate per cubic meter of about 23.6 watts [(2000+2000)*3.7E+10*1.6E-19].

On the other hand, the concentrated first cycle raffinates going into a typical commercial fuel reprocessing plant's glass melter would have roughly one mole per liter of fission products in it. Assuming that 10% of those fission products still behave like ^{137}Cs and ^{90}Sr (haven't decayed yet), the heat generated by 1 cubic meter (1000 liters) of such HLW would be about 7058 watts [1000 l/m^3*1 mole/l*6.023E+23 atoms/mole)*1.6E-19 J/ev *1E+6ev/atom)**ln2/(3600*24*365*30*] - that's fifteen thousand times « hotter » than is Hanford's HLW.

decay/s = Bq = number of radioactive atoms**ln2/half life in seconds*

(where ln2=0.693)

See https://en.wikipedia.org/wiki/Radioactive_decay

Appendix XII. How the Nuclear Industry's «experts» Sometimes Mislead Us

The purpose of this little rant is to demonstrate that understanding a procedure's technical details is important in determining whether or not what you are being told supports the teller's contentions.

During the time that INEEL served as DOE's lead radioactive waste management lab, the key criterion for evaluating candidate radioactive waste form materials was their performace on the «Product Consistency Test» (PCT). It was performed as follows:

- the sample is shattered and then ground to a powder
- particles between 75 and 150 microns in diameter are isolated by running powder through calibrated screens
- that fraction is subjected to a quick rinse with water or ethanol to rinse off fine dust after which it is dried (such rinsing was originally left to the discretion of the analyst)
- the rinsed/dried particles are weighed into into s Teflon or stainless steel «bomb» and 10 times as much deionized water is added
- The bomb is sealed and put into a 90°C oven for exactly one week
- Its contents (water leachate plus undissolved glass) are then filtered to isolate the liquid (leachate)
- That leachate is analyzed to determine the fraction of each of the sample's constituents solubilized (leached)

The pass/fail criterion for DOE's HLW glasses (both then and now) is whether a higher fraction of the sample's alkali metals (usually its most intrisically leachable/soluble major constituents) ends up in solution than does from a portion of its benchmark «Environmental Assessment» (EA) glass run through the same protocol. About 12% of EA glass's alkalies generally dissolve (EA glass is not a very durable glass – it's an «easy» standard).

The PCT is relatively quick/easy to perform and, if the sample in question is a glass (homogeneous frozen liquid), generates a genuinely meaningful result because everything within them is «released» at about the same concentration-normalized (fractional) rate; i.e., if 10% of the specimen's alkalies end up in solution, 10% of the radionuclides and chemical toxins have been «unsequestered» as well.

The rationale ginned up by DOE's experts to support its contention that « mineralized» steam reformer product would be «better than glass was that it is «... *primarily composed of nepheline (ideally* $NaAlSiO_4$*) and the sodalite family of minerals (ideally* $Na_8[AlSiO_4]_6(Cl)_2$ *which includes nosean (ideally* $Na_8[AlSiO_4]_6SO_4$*). Semi-volatile oxyanions such as* ReO_4^-, TcO_4^-, *are expected to replace sulfate in the larger cage structured nosean and halides such as* I^- *and* F^- *are expected to replace chlorine in the nosean-sodalite mineral structures – thereby immobilizing them.*»

In reality that's not the case. When INL's SBW simulants were «reformed» the majority of the stuff going into the «product» receivers was «fines» consisting of <10 micron-sized, poorly mineralized dust mixed with lampblack-like elemental carbon particles. X ray diffraction spectrometry (XRD), the technique utilized for characterizing the minerology of such things, is «blind» to much of that stuff because it's not very crystalline and XRD doesn't identify what's within its crystalline moiety (e.g. chlorine, ^{99}Tc, or sulfate) - only its structures (XRD is a qualitative not quantitative analytical method).

The PCT protocol's 's 75 micron lower particle size cut-off dictates that the analyst deliberately not test ~2/3's of what most such steam reforming's demonstrations actually produced – its relatively poor-performing fines product fraction. That fraction contained the bulk of the waste's intrinsically volatile constituents (e.g., cesium, iodine, chlorine, sulfur, technetium, cadmium, etc.) remaining in either of its solid products and also tends to be more water soluble (leachable) than the coarser sand-like bed product particles [183].

Furthermore, almost all of the solids produced by a steam reformer ares multiphasic which means that their constituent elements are present in different forms exhibiting different characteristics - not dissolved within a homogeneous frozen liquid (glass). This, in turn, means that literal adherence to the formal PCT protocol is apt to generate wildly optimistic (misleading) results. For example, a quick water leach of either of the products of INEEL's (fines and its «grapenuts like» granular product) steam reforming demonstrations removes most of their chloride [183]. Since the original version of ASTM C1285-02 (the one used during the crucial initial decision-making phase of the project) left sample powder-washing to the discretion of the analyst, an "official" test performed for the purpose of determining the leach rate of chloride could be done on material that no longer actually contained chloride - which, in turn, means that the final analysis's apparently very low Cl^- result might be (and was) interpreted to mean that steam reforming had immobilized it within a leach resistant tectosilicate mineral (sodalite, $Na_8Al_6Si_6O_{24}\ Cl_2$). The same rationale supported claims that other tough-to-vitrify species (e.g., sulfur and ^{99}Tc) would end up within another leach resistant «cage mineral» (nosean).

Additional deliberate confusion was generated by normalizing the PCT's raw "fraction leached" results to bogus surface area.

For example, one gram of cubic mid range-sized, 2.6 g/cc PCT sample particles (diameter =0.01125 cm [0.015 +0.0075)/2]), would have a geometric surface area of 205 cm^2 [6*0.01125^2/0.01125^3/2.6]. However since a steam reformer's solid products are intrinsically porous, their BET surface areas are far larger than their geometric surface areas[237].

In a typical such leach test that I'd performed myself [183], 12.5% of the sodium in a 50:50 mix of INEEL's pilot plant steam reformer's fines & bed products dissolved. If that figure is normalized to its «geometric» surface area, its area normalized leach rate in the usual units works out to 6.1 g/m^2/week [0.125/(205/10000 cm^2/m^2)], or about the same as that of DOE's go/no-go standard glass. However, to make the steam reformer product's granulated product fraction seem superior, DOE's leach test experts at both SRS and Hanford decided to normalize fraction-leached PCT results to a "total" (BET) sample particle area[238] rather than upon a figure based upon particle size and density as the test's protocol dictates [240].

Reporting such «normalized» results made that stuff appear to be >100 times more leach resistant than DOE's benchmark glass and therefore gave

---

[237] The BET surface area of a 50-50 by weight mix of the as-generated fines and bed products produced during the "optimized" demonstration described in APPENDIX IV was 83m2/g. The samples subsequently characterized by SRS were first subjected to an overnight "cook" at 500°C which would almost certainly serve to reduce BET-type surface areas as well as burn off the "coal". Nevertheless, they still exhibited BET surface areas over a hundred times greater than would the same sized glass particles.

[238] The BET measures the amount of an inert gas condensing upon on a powdered sample at a temperature near the boiling point of that gas. The amount so condensed is measured by the pressure reduction within the system when sample is introduced. All open pores, inclusions, irregularities, etc., penetrable by the inert gas (usually nitrogen) are accounted for which means that anything that's intrinsically porous (e.g., charcoal, silica gel, or FBSR calcines) will exhibit a far larger BET than geometric surface area.

great reassurance to those analysts' decision-maker customers. However, it's also transparently(?) silly because any material possessing a surface area/gram greater than about 1640 cm^2/g (205*1/0.125) could totally dissolve during the PCT and still be characterized as "better than glass" (this little technical detail was never pointed out (realized?) by the experts responsible for writing/reviewing such reports[239]).

Even more important, it is unreasonable to characterize a "mineralized" (or any other) calcine as superior to glass (or even to a properly made concrete) based solely on PCT results – even results obtained with a "conservative" version of it [183]. The reason for this is that glass is intrinsically monolithic and calcine is intrinsically dust-like. Since real world leaching occurs at « outside » surfaces, the surface area of the *intact* waste form is what would count in a repository. For example, a one million-gram glass monolith (smaller than most real or proposed US glass waste forms) possesses a geometric surface area of about 3 m^2. Its "real" surface area isn't much greater than that because glass is basically just a non porous super-viscous liquid [241]. About one gram (exact amount depends upon how its area is measured) of INEEL's mineralized FBSR test products possessed that much surface area.

[239] A reason for this was identified by Upton Sinclair about a century ago: "*It is difficult to get a man to understand something, when his salary depends upon his not understanding it*".

Appendix XIII. Example of a Promising Concept that Needs Experimental Verification as Soon as Possible

The data in Table 7 (below) is excerpted from ORNL 2751 [116]. Its numbers were based upon calculations performed with a 31-group, multiregion, spherically symmetric, diffusion code UNIVAC program named "OCUSOL". Group-averaged cross sections for the elements of interest were based on then available data. Where such data were lacking, reasonable interpolations based upon resonance theory were used. Estimated neutron reaction cross sections were made to agree with measured

Table 7 ORNL 2751's Case 35 Spherical "clean core" two salt thorium breeder.

Core diameter ft	3.0	Neutron absorption ratios continued	
Mole% ThF_4 in fuel salt	0	Core vessel**	0.0902
Mole% ^{233}U in fuel salt	0.592	L i and F in blanket salt	0.0233
^{233}U atoms/cc of fuel salt x10^19	21.1	Thorium in blanket salt	0.9722
Neutron absorption ratios*		Leakage***	0.0477
^{233}U(fissions)	0.8754	Core volume	400 liters
^{233}U capture (n,y)	0.1246		
Be, Li, and F in fuel salt	0.0639		

*ORNL assumed 2.1973 neutrons/fission (='s a total absorbed per neutron absorbed by ^{233}U)

**ORNL assumed a one third inch thick INOR 8 core tank

***ORNL assumed a two foot-thick, 25 mole% ThF_4/75mole% ^7Li F, blanket salt

resonance integrals where available and saturations and Doppler broadening of the resonances in thorium as a function of concentration were estimated.

The reasons why tests verifying the conclusions of this particular set of ORNL's calculations should receive immediate attention include:

- If Weinberg's team was right, then the full-sized tube-in-shell concept that I've described (Fig. 14) is very likely to behave as predicted and therefore represent the "best" way to implement a sustainable nuclear renaissance
- It would be a relatively cheap system to both build and operate (much cheaper/simpler than any sort of solid fueled test reactor)[240]
- Its operation would enable the generation of real data[241] relevant to every aspect of both building and operating breeding capable MSRs
- Systems like it could be readily scaled up by substituting one or more cylindrical cores for the single spherical core contained within its blanket salt tank

The reasons why it should be at least as capable as ORNL's modelers predicted include:

- It seems that ORNL underestimated the number of neutrons generated per ^{233}U fission (i.e. >2.3 rather than 2.2) which

---

[240] For instance, its core would require only about 32.6 kg [400 liters*1000 cc/liter*21.1E+19/6.023E+23*0.233] of ^{233}U fissile to achieve criticality (operate). Depending upon how much heat is to be generated/dissipated during its testing, that figure would be somewhat bigger. For example, if we assume that 9 cubic meters of outside core volume is sufficient to remove 3 GWt (the figures estimated for the EU's MSFR), a 50 MW test reactor would require 150 liters of additional core salt boosting its total fissile requirement to 45.6 kg. Similarly the Hastelloy N "supermetal" required to make a three foot diameter, one quarter inch thick spherical core tank would weigh about 637 kg and cost about $16,000 (see https://www.alibaba.com/showroom/hastelloy-n.html). If 316 stainless steel (SS) were to be substituted for it instead, its core tank would cost about one fifth that much. With proper redox control, 316 SS is probably good enough.

[241] Its salt streams would provide something real for the folks working out any fluoride salt MSR concept's fuel clean up and waste management schemes to work with. Its behavior would provide updated neutronics data enabling better prediction of that of a full sized system. Finally, realistic corrosion/neutron damage rate data would be generated – almost 100% of the MSR-relevant recent corrosion testing has been done "cold" (no neutrons) with pure salts containing no fission product surrogates.

would boost Table 7's predicted CR (0.9722) to well over 1.0 [105]

- There's no compelling reason to assume that a 3 foot diameter core tank immersed in a blanket salt would would have to be one-third inch thick – a thinner core tank wall would also boost CR
- There's also no compelling reason why the blanket salt tank couldn't be considerably bigger which would reduce neutron leakage and therefore again boost CR

Appendix XIV. INL's Steam Reforming Process

During the 1990's, a Swedish firm, STUDSVIK, developed a process to "burn" the organic wastes generated by civilian nuclear fuel cycles (mostly the ion exchange resins utilized to purify LWR coolant water) in a way that avoided the negative connotations that had come to be associated with "incineration". It utilized superheated steam (hence "steam reforming") to decompose such materials to simpler molecules (methane, hydrogen, carbon monoxide, etc.) under the conditions used to convert "biomass" to gaseous fuels. When it became apparent that US DOE wanted a substitute for calcination too, the same(?) process was suddenly construed to be the "best" way to treat salt-type wastes as well. Consequently, in 2002 Studsvik joined with Westinghouse Government Environmental Services Company LLC to form a new company, THOR Treatment Technologies to further develop, promote and deploy that technology (see [242]).

The resulting fluidized bed steam reformation (FBSR) process possessed four key virtures: 1) its "Denitration and Mineralization Steam Reformer" (DMR) would not directly oxidize organic components of such waste streams with elemental oxygen, hence it's not waste "incineration" (real DOE tank wastes contain little or no organic matter); 2) since the immediate solid products produced under its strongly reducing (fuel>>oxygen) reaction conditions were mostly carbonates, not oxides, it wasn't "calcination" either[242]; 3) since alkali nitrate salts should decompose under such conditions (eventually anyway), it's supposed to prevent bed agglomeration and simultaneously eliminate the offensive plume of NOx emitted by INEEL's traditional approach to radwaste calcination (the same things that sugar calcination would have accomplished); 4) the inclusion of powdered clay

[242] This is another example of something being characterized in terms of what it's not rather than in terms of what it is. Reforming is nicer sounding too – Norway "reforms" its murderers instead of "frying" (calcining?) them in electric chairs.

along with the solid fuel granules[243] employed to provide reaction heat[244], serve as a reductant, and generate a clean-burning[245] intermediate offgas, was supposed to convert SBW to a poorly soluble "Grapenuts-like" mineralized product purported to be a "better than glass" disposal form; and 5) several of the especially troublesome (to glass makers) anionic components of DOE's salt wastes (e.g., sulfate and chloride) were also supposed to be simultaneously sequestered within leach resistant aluminosilicate "cage minerals".

The following pages (smaller font) are excerpted from a report that I wrote/presented in 2004 about experiments performed to determine whether the product(s) of a local "demonstration" of THORT's technology could be rendered "nondispersable" with cements [243]. There's an error in its second footnote: I subsequently discovered that THOR's "open access" technical reports had misrepresented both its subcontractor's (HAZEN's) experimental system and its products – fines were apparently never successfully 100% recycled and thereby just one, "grapenuts like" product produced as repeatedly claimed.

During Dec. 2001 several tests of Studsvik's "mineralization" process were performed by Hazen Research, Inc, of Golden CO. 570 liters of a concentrated (~44% solids) aqueous salt solution representing a typical "low level" Hanford tank waste were processed in a six-inch diameter reactor. The primary component of this simulant was ~8 molar sodium ion balanced by (in order of decreasing concentration) hydroxide, nitrate, carbonate, nitrite, carbonate, aluminate, sulfate, fluoride, phosphate, and chloride. It also contained lesser amounts of several other metals plus about 80 grams per liter of an assortment of the organic chelating agents used in Hanfords' processes. The reactor's alumina bed particles were fluidized with superheated steam and ground coal was added to create a strongly reducing environment. Sufficient kaolin (nominally $Na_4Al_4Si_4O_{10}.8OH$) and powdered quartz were slurried with the waste simulant to produce a product with roughly the same gross composition as nepheline, $NaAlSiO_4$. Samples of it were submitted to the Savannah River Technology Center which subsequently issued a report supporting Studsvik's claims and recommending that DOE pursue the technology.

---

243 (coal, charcoal, etc. –a solid reductant constitutes the "novel" feature of THOR's patent).

244 Sufficient oxygen was added to burn enough of the coal (certainly not all of it) to keep the DMR's reactor hot enough to function.

245 The final version of THORT's steam reformer (THOR no date) differs in how its DMR's off gas is "burned". After most of the dust is removed via cyclones & blow back filters and sent to a "product" container, the DMR's gaseous products (mostly steam along with combustible hydrogen, ammonia, carbon monoxide, hydrogen, cyanogen, etc.) passes into a second even hotter (~1000°C) "Carbon Reduction Reformer" (CRR - another fluidized bed reactor) to which sufficient additional oxygen is added to burn it to a mixture mostly comprised of elemental nitrogen, excess oxygen, carbon dioxide, and water vapor.

In January 2003, a demonstration of how it could deal with a representative INEEL liquid SBW (Table 1) was performed at the Science Applications International Corporation (SAIC) Science and Technology Application Research (STAR) center located at Idaho Falls, Idaho. This test was performed under conditions mutually agreed upon by THOR and INEEL personnel and, while it was subsequently deemed "successful", it was also agreed that there was a good deal of room for improvement. Consequently, in November 2003 a two week-long "optimization run" was run at the same facility. This paper describes what was produced while the reactor was being run with a mineralization flowsheet (i.e., with clay added to the SBW) and how cementitious technologies could convert this material into a superior waste form.

Experimental: Figure 1 depicts the FBSR reactor used for SAIC/INEEL's most recent FBSR test. It basically consists of a 30 inch length of six-inch internal diameter pipe reaction zone situated under a five foot-long, expanded (12 inch diameter) particle disengaging (or "freeboard") zone. Both sections are made of an especially corrosion resistant INCONEL alloy. Table II lists typical operating conditions. One of the reactor's two feed streams consisted of an aqueous slurry comprised of various proportions of sugar-syrup with Table I's SBW simulant plus kaolin. This cream-like

Table 1 INEEL tank WN 180 SBW simulant.

CATIONS			
	Molar/grams/l		**Molar/grams/l**
Acid (H+)	1.1/1.13	Magnesium	0.012/0.292
Aluminum	0.66/17.9	Manganese	0.016/0.775
Boron	0.012/0.133	Mercury	0.0014/0.271
Calcium	0.047/1.89	Nickel	0.0013/0.086
Cesium	0.0032/0.431	Potassium	0.20/7.66
Chromium	0.0034/0.174	Rhenium	0.0011/0.20
Copper	0.0007/0.044	Sodium	2.1/47.4
Iron	0.022/1.21	Strontium	0.0020/0.176
Lead	0.0013/0.274	Zinc	0.0011/0.069
ANIONS			
Chloride	0.03/1.06	Phosphate	0.029/2.74
Fluoride	0.047/0.90	Sulfate	0.070/6.72
Nitrate	5.3/330	Water	40.6 (731g/l)

slurry was sprayed into the lower end of the reaction zone with a pneumatic nebulizer utilizing nitrogen as the motive gas. The other feed stream consisted of coarse granules of activated charcoal plus occasion additions of an iron oxide "de-NOX catalyst". The external surfaces of the reactor & its disengaging head were covered with clam-shell electrical heaters. A distributor plate situated immediately above the flat bottom of the reaction zone introduced superheated steam which served both to fluidize the bed particles and provide a portion of the heat required for the strongly endothermic "reforming" reactions. Any additional heat required to maintain the system at the desired temperature was provided by "reforming" elemental oxygen added to this steam. A cyclone served as the primary collection system for the "fines" (comprised of small particles of the reactor's product plus elemental carbon dust) elutriated from the reactor. A screw auger continuously recycled that dust back into the base of the reactor[246]. Particles too small to be captured by the cyclone were collected by a bank of sintered metal blow-back filters situated within a cyclone-shaped housing. This portion of the reactor's "fines" plus the material retained in the bed constitute its two product streams[247].

Typical FBSR reaction conditions

- Gases: 7 kg/hr steam + 3 kg/hr N_2+1 kg/hr. O_2
- Liquid Feed: 4.5 kg/hr SBW/clay slurry[248*] + 3 kg/hr of 55wt% sucrose syrup
- Solid Feed: varying amounts of coarse activated charcoal** + occasional additions of FeO_x
- Temperature: ~720°C

Filtered offgas was diluted with propane plus excess air and passed into a ceramic-lined thermal oxidizer large enough to provide a mean gas residence time of approximately one second at 1000°C. After the temperature of the burner's offgas was lowered to roughly 300°C by spraying water into it, it was sucked through a venturi scrubber into the scrub tank. The quenched/scrubbed gas exiting that tank was passed through a demister,

[246] Occasionally this auger would be reversed in order to collect samples of the cyclone catch.

[247] The fact that this portion of the reactor's fines were not recycled to the reaction zone constitutes the most significant difference between the HAZEN and SAIC/INEEL tests – HAZEN produced only a "bed" product.

[248] 290 grams of kaolin were added per liter of SBW: this much clay increases the volume of the liquid by 10% and its SpG from ~1.25 to ~1.38. The sugar feed rate varied considerably but was always well in excess of that required for stoichiometric reduction of nitrate to elemental nitrogen.

reheated to roughly 120°C to prevent condensation, sucked through a three-stage bed of activated charcoal, and, finally, blown through HEPA filters to the stack.

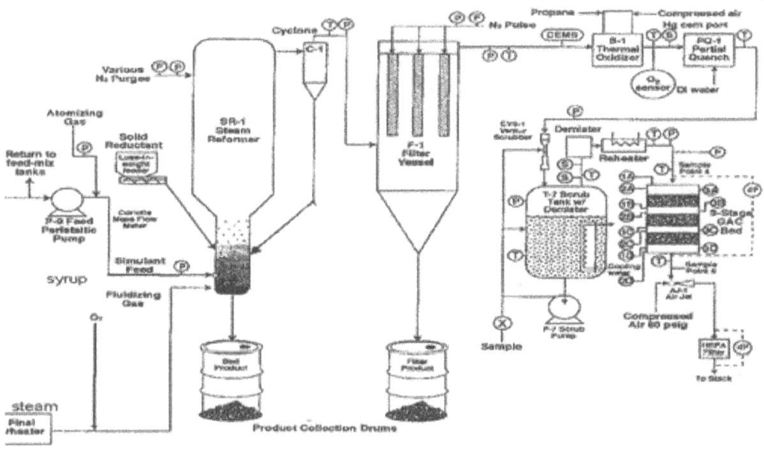

Since the primary purpose of this paper is to discuss "monolithification" of the reactor's product, its description of how the "reformer" itself performed will be limited to the following observations:

1. The processing of sufficient SBW/clay slurry (498 kg) to generate 126 kg of mineralized product produced a total of approximately 74 kg of "fines product", 3.6 kg of cyclone dust samples, and 44 kg of "bed product"[249]
2. On-the-fly samples of both products during the run generally contained a substantial amounts of elemental carbon (dust in the fines fraction, 1-5 cm chunks in bed samples)– the amounts of "product" carbon depended upon the feed rate of activated charcoal to the reactor
3. Bed product samples generally contained mineralized agglomerates[250] which tended to grow larger and more numerous as the run continued

[249] This figure (44 kg) does not count the ~52 kg of ~0.5 mm-diameter, sintered alumina "starting bed" charged to the reactor on two separate occasions.

[250] The water solubility and carbonate concentration of these lumps (both low) were similar to those of the non agglomerated bed material.

4. On two (three?) occasions those agglomerates grew large enough to cause bed defluidization which immediately shut the process down
5. Regardless of charcoal feed rate, "on the fly" samples of either solid product fraction never contained more than 0.1% residual nitrate (or nitrite) – generally none at all
6. Offgas samples taken upstream of the offgas burner generally contained several thousand ppm (by volume) ammonia plus readily detectable concentrations of HCN. The ammonia concentration correlated directly with charcoal feed rate[251]
7. Total NO_x (NO_2+NO) concentrations at the same point was generally somewhat higher than that of ammonia and inversely related to charcoal feed rate
8. The scrub liquor was always strongly acidic – pH generally less than 2.5 – and contained very high concentrations of sulfate, phosphate, chloride, and cesium relative to iron, sodium, aluminum, etc.

These observations support the following conclusions:

1. FBSR tends to volatilize more of a radwaste's semi-volatile components than does rotary kiln calcination
2. FBSR's lessened NO_x production comes at the price of producing much more of such reduced nitrogenous species as ammonia and hydrogen cyanide – any kind of calciner is going to need subsequent offgas treatment!
3. There little evidence that the "sodalite-like" cage minerals purported to immobilize anions such as chloride actually formed

FBSR product characteristics: The following is based upon analyses of samples of what was in the two drums into which filter fines and bed products were dumped throughout the duration of the test.

- Ten-minute exposure of either powder to 100-fold as much 90°C water solubilized only a small fraction of the

[251] The reasons for this are 1) fluidized bed reactors require higher gas flow rates; 2) the temperature required to make "water gas" are two hundred degrees higher than is required to efficiently sugar calcine radwaste; and 3) the vapor pressure of a molten metal is generally greater that that of its oxide.

sodium - about 20% of that in the fines fraction, about 4% of that in bed product. This suggests that FBSR does indeed "mineralize" most of the sodium.

- The primary cation present in water leachates was sodium balanced by an equivalent amount of (in order of concentration) aluminate, carbonate (in fines, not bed), phosphate, sulfate, chloride, silicate, and fluoride ions.
- The bulk density of the filter fines product fraction was about 0.35g/cc[252] - that of the bed product, ~0.8 g/cc.
- Several intrinsically volatile components of the SBW simulant – rhenium (technetium surrogate), cesium, chloride, sulfate, and phosphate – were present at much higher (several times as high) concentrations in the filter fines product fraction than in the bed product
- The sum of all of sulfur-bearing species (sulfate, sulfite, and sulfide) found in both product fractions (plus the scrub liquor) accounted for less than one-third of the sulfate-sulfur fed to the reactor
- The loss-on-ignition (primarily elemental carbon) of the final bed product composite was ~4.5% vs ~15 wt% in the filter fines drum
- There was under 1 ppm of either ammonium or cyanide ion in either product
- Hot water would dissolve the same amount of chloride from either product fraction as would high temperature fusions performed with NaOH or Na_2CO_3

The overall degree of "volume reduction" achieved by "reforming" this SBW was not very impressive. For example, if we assume that …

1) a total of 385 grams of mineralized product is produced per liter of SBW
2) the mass-wise proportions of bed and fines products is 40:60
3) "Bed" is 5 wt% carbon & "fines" is 15% carbon
4) the bulk density of these fractions are 0.8 and 0.35 g/cc

[252] The "bulk density" of a powder is not an absolute number because it depends upon its degree of consolidation (settling) before its volume is measured. The figures given in this paper were obtained by dumping powder into a tared 10 cc glass graduated cylinder, tapping it for about 30 seconds, and then measuring its mass and volume. That much tapping typically reduces the volume of a dumped "fines" sample by about 50%.

a reasonable estimate of the volume of calcine generated from one liter of SBW would be ...

$$((0.6x385)/0.85)/0.35 + ((0.4x385/0.95)/0.8 = 979 \text{ cm}^3$$

To a "grout" chemist, both product fractions were excellent from a purely chemical point of view and rather poor from a physical standpoint. "Good" because a nepheline-like mineral assemblage is intrinsically compatible with silicate-based cements. "Poor" because their low bulk densities and high BET surface areas causes them to exhibit high "water demand" (fines more so than bed). In practical terms, this means that a large amount of liquid is required to produce a grout that would flow well enough to readily fill a mold (or canister). Since the total pore volume of fully-cured grout (concrete) generally approximates that of the water in the original formulation, this in turn, means that "high water demand" translates to a physically weak, porous, concrete. It also translates to low waste loading, both weight and volume-wise.

Due to time (and funding) constraints, only four fundamentally different types of grout were investigated and the total number of specimens produced was twenty-one. The waste simulant in these specimens generally consisted of a 50:50 weight-wise mix of bed and filter fines. Similarly, the weight-wise percentage waste loading was also usually 50% (i.e., the mass of dry cementitious agent(s) equaled that of the calcine).

Etc., etc

A comparison of the observations that others made during this demonstration (also see [244]) with how THOR's reports characterized it (see [242]) differ in many ways, the most important of which (to me anyway) is that THOR doesn't bother to mention that only about one third of the waste surrogate's primary ash forming constituents (sodium, aluminum, etc.) ended up in its much ballyhooed mineralized "grapenuts-like" product – the majority ended up as fines. They also don't reveal that much of the surrogate SBW's intrinsically volatile stuff (e.g., chlorine and sulfur) didn't end up in either solid product – such things were mostly converted to gases, not sequestered within durable "cage minerals" as THOR's DOE Complex "helpers" often suggested. More importantly (to me anyway), the reports written by the national laboratory personnel that chose to "help" THOR didn't point those things out either.

To date (4April2019), the roughly one billion dollars subsequently spent by INL's clean-up contractors on its steam reformer hasn't yet succeeded in converting any of its remaining SBW to either a "carbonate" or "mineralized" product. The problem seems to be the same one that plagued us during the "demonstrations" performed at INEEL's STAR center almost two decades ago, fluidized bed agglomeration and "bark" formation. The chunky stuff in question is comprised of the DMR's fluidized bed particles glued together with waste surrogate-derived sodium/potassium carbonate. The much higher temperatures (~700°C) required to denitrate such waste with THOR's proprietary solid reductants rather than with a water soluble reductant (~500°C) are responsible for this. With something like sugar, the reductant is intimately mixed with what it's to reduce (the liquid waste's nitrate ions), not in an entirely separate phase, and therefore can completely react at temperatures well below those likely to melt/fuse the product.

APPENDIX XII describes one of the ways used to "sell" steam reforming to DOE's stakeholders.

Index

Lightning Source UK Ltd.
Milton Keynes UK
UKHW022218230919
350308UK00003B/97/P

9 781119 657781